German Tradenames and Trademarks 1900-1945

An Encyclopedia of German Tradenames & Trademarks 1900-1945

by W. Darrin Weaver, MPA

Schiffer Military History
Atglen, PA

Book Design by Ian Robertson.

Library of Congress Control Number: 2011920799

Printed in China.
ISBN: 978-0-7643-3759-8

We are interested in hearing from authors with book ideas on related topics.

Published by Schiffer Publishing Ltd.
4880 Lower Valley Road
Atglen, PA 19310
Phone: (610) 593-1777
FAX: (610) 593-2002
E-mail: Info@schifferbooks.com.
Visit our web site at: www.schifferbooks.com
Please write for a free catalog.
This book may be purchased from the publisher.
Please include $5.00 postage.
Try your bookstore first.

In Europe, Schiffer books are distributed by:
Bushwood Books
6 Marksbury Avenue
Kew Gardens
Surrey TW9 4JF, England
Phone: 44 (0) 20 8392-8585
FAX: 44 (0) 20 8392-9876
E-mail: Info@bushwoodbooks.co.uk.
Visit our website at: www.bushwoodbooks.co.uk

Contents

Introduction

I assembled this reference for collectors, researchers, other authors and dealers as an adjunct to a number of books already in print. If you purchased or opened up this book, chances are you already know a thing or two about German firearms, militaria, cutlery or optics, in addition to the markings, codes, and stamps associated with them. A number of references have already been published over the years covering each of these topics. For instance, several outstanding books exist which describe the WWII-era military coding system that started with letter-number combinations and evolved to "a" to "ozz" and beyond; my friend Michael Heidler's *Deutsche Fertigungskennzeichen bis 1945* is the best and most comprehensive of these efforts. Joachim Görtz and Don L. Bryans did a remarkable job in documenting small-arms unit marks in *German Small Arms Markings*, and other books have been produced over the years that cover proof and acceptance marks. LTC (Ret.) Thomas Johnson's extensive research into the German cutlery industry has provided historians and collectors a number of invaluable references which describe the codes and markings found on knives, swords, and daggers of the era, and many excellent published and open-source references exist regarding the Leistungs Gemeinschaft der Deutscher Ordenshersteller (LDO) and Reichszeugmeisteri (RZM) marking systems. In addition, all of my previous works covering firearms include obligatory sections on codes and markings, and *Kunststoffe: A Collector's Guide to German WWII Plastics and their Markings* is almost completely devoted to the subject of plastics manufacturers, codes and markings.

As I collected, studied and researched firearms and militaria over the years, though, I frequently came across interesting, unidentified, or unfamiliar logos and trademarks. No single reference then existed to help determine their meaning, origin, or ownership. Thus, as other projects evolved, I at least tried to gather as many documents, notes, photos, and leads to eventually produce what I hope you will find to be a handy, fairly broad reference covering German small-arms, optics and cutlery tradenames, trademarks, and logos from the era.[1]

It is impossible, of course, to assemble any one reference that lists every single German tradename, trademark or logo that one may encounter between 1900-1945, forty-five years being a pretty good span of time and the entire German economy mobilized for war production during WWI and WWII. Millions of articles were produced by scores of companies that went into and out of business. Brands, tradenames, trademarks and logos were frequently transferred, sold, resurrected, or simply died-out during this period, complicating matters, and thus this book should be by no means considered complete. It should serve, however, as a reliable starting point if one discovers an item with an unfamiliar name or marking.

Of special note, many of the names and products discussed in this work are still proprietary and protected by international trademark laws and agreements. Some have become colloquial generics over time, the status of others is unclear, and many of the original trademark permits have expired. Trademarks and logos have been reproduced solely for reference and identification and no infringement is intended. Most images of trademarks were obtained from public records on file at either the United States Patent and Trademark Office or the Deutsches Patent- und Markenamt. Those that could not be found there were recreated digitally by the author.

Images used and reproduced by permission are credited in the text. The remainder of the images not directly credited were photographed by myself off items in my personal collection, or were obtained from my collection of wartime press and periodical images, all of which are thought to be public domain.

Lastly, I wish to extend my gratitude to the many contributors who helped with this project, and especially my friends and fellow authors Michael Heidler, Kenneth Huddle, Robert Jensen, Paul Shomper, Jon Speed, and Dr. Geoffrey Sturgess, all of whom kindly contributed information, photos, proof read, and corrected my work along the way. A special thanks to my family, which endlessly suffers because of time away researching and writing.

I thank you for purchasing this book and I hope that you find it interesting and useful.

W. Darrin Weaver
2010

Contributors and Contributing Companies and Institutions

Contributors

Lance O. Adams
Jan Balcar
Joseph Bielik
Ed Buffaloe
Ron Clarin
John Cross
Bob Cutler
CSM (Ret.) David W. Davis
Leon DeSpain
Reiner Eckert
Bob Faris
Werner Frank
Michael Heidler
Kenneth Huddle
Robert Jensen
Bob Marquard
Dieter H. Marschall
CWO4 Duward Massey II
Nick Messer
Bernd Niebling
Warren Odegaard (†)
Paul Philipp Pöllinger
Walter Schmid (†)
Paul Shomper
Brad Simpson
Robert Simpson
William R. Smith (†)
Jon Speed
R. Blake Stevens
Dr. Geoffrey Sturgess
Zuzana Veselá
Paul Viens
Jason C. Weaver
Margaret J. Weaver
Mike Wieringa

Contributing Companies and Institutions
Bayonet Collector's Network
Bundesarchiv, Koblenz, Germany
Collector's Firearms, Houston, Texas
Collector Grade Publications, Cobourg, Ontario
Deutsches Patent- und Markenamt, München, Germany
Historical Parts and Militaria Co., Eddy, Texas
Landesarchiv Nordrhein-Westfalen, Germany
Manion's International Auction House, Inc., Kansas City, Kansas
Meopta-optika, s.r.o., Přerov, Czech Republic
Military Rifle Journal, Snowflake, Arizona
Ministry of Defense Pattern Room, United Kingdom
National Rifle Association, Arlington, Virginia
City of Oberndorf Museum, Oberndorf, Germany
United States Army
United States Army Ordnance Museum, Aberdeen Proving Ground, Maryland
United States Department of Veteran's Affairs
United States National Archives and Records Administration, College Park, Maryland
United States Patent and Trademark Office, Department of Commerce, Washington D.C.

1

A Brief History of Branding and Trademarks

The marketing business has been around for many centuries, to the point that today the study and development of brands, brandnames, and trademarks is almost a science unto itself. Marketing and branding, however, really came into widespread use in the 19th century with the dawn of industrialization, and particularly the advent of packaged goods.

Cattlemen had used red-hot irons to burn their initials or other identifying marks into the hides of animals for generations. These "brands" were unique to the individual or company, and were necessary in the age of "open range" grazing to determine just whose cows were whose. Gradually, in order to help solve duplication issues and other disputes, a system of registering these marks at the nearby courthouse became widespread. This, of course, is where the marketing terms "brand" and "branding" originated.[2]

By the early 1800s, many goods and items that had traditionally been made locally or in the home were being made in far-away factories. Most of these firms shipped their wares in wooden barrels or boxes which were not disposable, but rather, were meant to be returned to the company for re-use. To help ensure that the shipping containers were returned to their appropriate owners companies, like their cowboy predecessors, began literally "branding" their company names and addresses into the wood. Other companies developed and began using distinctive insignia or logos, which were in turn "branded" or marked into the product and/or the shipping container, and these evolved and in turn became known as "trademarks."

Consumers gradually became used to seeing mass-produced items arrive from afar bearing a certain name, logo, or marking, and many became fond of one manufacturer over another. Some companies and products became so well known and respected that their names became "brand names." The manufacturers, for their part, began to recognize the way in which consumers were developing relationships with their brands, or "brand loyalty," and the marketing value behind these names, as well as their logos and other proprietary markings. Before long others began cashing in on established brands by producing "knock-offs," or otherwise imitating or mimicking respected names and markings. This gave rise to companies wanting to legally protect their brandnames, trademarks, and logos, which in turn led to the rise of national registration systems.

It was much the same in Imperial and National Socialist Germany. Indeed, most German manufacturers etched, branded, stamped, painted or otherwise marked their proprietary tradenames, logos or trademarks into the products they manufactured long before 1900. Manufacturers of firearms, optics and edgeware were no different, and also boldly marked their names, logos and trademarks into the items they produced up to, through WWI, and beyond.

The *Deutsches Markenamt* was set up to centrally approve and catalog the names and marks. Manufacturers registered all sorts of combinations: abbreviations, initials, names, brands, locations, and all sorts of figurative elements. For example, Fritz Walther of Waffenfabrik Walther not only trademarked his family name, but also the company's pennant, or "banner." He even claimed and was granted ownership of a depiction of the flowing banner itself.

Fig. 1: The Waffenfabrik Walther flowing pennant trademark as registered with the *Deutsches Markenamt*.

Fig. 2: Walther even went so far as to register the figurative element on its own, the pennant itself, which is referred to as a "banner" by collectors.

2

Development of Regional Industries, Identities and Reputations in Germany

Just as certain companies developed their specific brands and reputations, so too did various regions in Germany. In fact, some regions became so specialized or well known for a particular industry that their regional or municipal names themselves became registered brandnames.

Why certain regions developed certain industries and industrial cultures has as much to do with the transportation, labor, technical, and natural resources available in one location over another as much as anything else. These same regions, and many towns and businesses within them, eventually began to recognize the value of their geographical reputations and incorporated them into their individual tradenames, trademarks, logos and advertising. A short review of these areas which are pertinent to this reference follows.

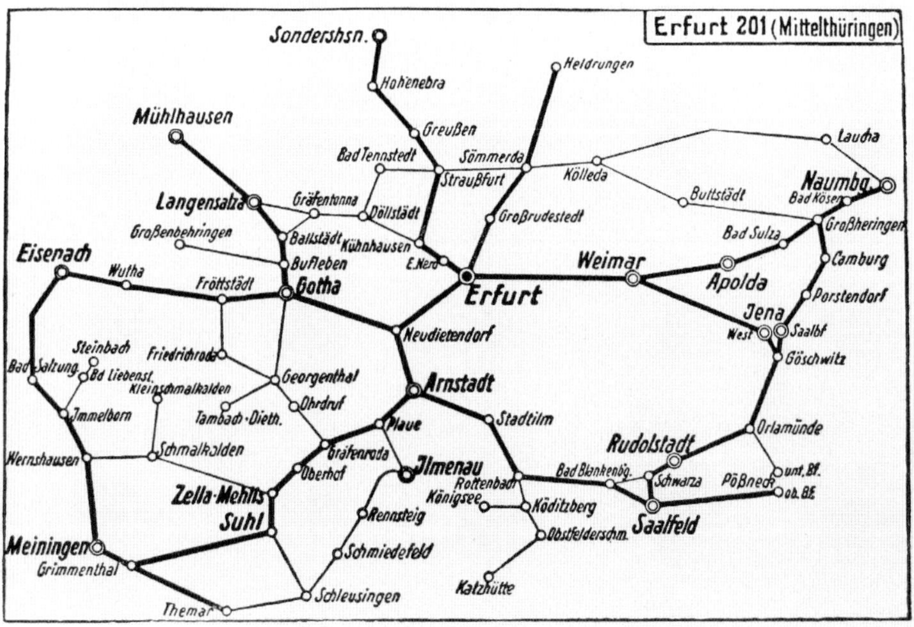

Fig. 3: A period map of the Thuringian arms-making region.

Small Arms

Thüringen has always been the center of the German small arms industry, with Suhl serving as the hub of the trade for many centuries.[3] The region was and is ideally suited for the trade, with abundant water power to drive furnace blowers, hammer forges, and grinders. In addition, the region is blessed with abundant iron ore in the surrounding mountains, with plentiful hardwood forests to fuel the forges and create grips and stocks. Lastly, it is centrally located and convenient to a number of traditional trade and transportation routes.

The town of Suhl coined itself as *the* German "Waffenstadt," and a plethora of family-owned and corporate gunmakers were first established in and around the area. Merkel, J.P. Sauer, Heinrich Krieghoff, Simson/BSW/Gustloff, Bergmann, and C.G. Haenel were the best known, but other smaller gunmakers operated in and around the town. These include Ziegenhahn, Heym, Bernhard Merkel, Gebhard Merkel, Oskar Merkel, Franz Jäger & Co., Hubertus, Griefelt, Nimrod, Paul Meffert, Schilling, Greifelt & Schlegelmilch, Fritz Kieß & Co., Gebrüder Adamy, H.D. Lindner, Eduard Kettner, August Menz, Emil Kerner, August Schüler, Walter Munk, Franz Neumann, Schmidt & Habermann, Gebr. Walter Rempt, Willi Barhold, Alfred Funk, A. W. Wolf, Immanuel Meffert, Thieme & Schlegelmilch, Johann Wilhelm Spangenberg, Wilhelm Gottlieb Spangenberg, and after WWII the Ernst Thälmann Werke (DDR), to name just a few, were proud to incorporate the name "Suhl" into their brands, tradenames, trademarks and advertising.

Nearby Zella-St.Blasii, which was later renamed Zella-Mehlis, also spawned its share of famous and not so famous gunmakers. These include Carl Walther Waffenfabrik, Adler, Fr. Langenhan and J.G. Anschütz (JGA), Anschütz, Bader, Barthelmes, Böhner, Bolte & Anschütz, Bosler, Büchel, Capell, Diem, Diemar, Ditzel, Döll, Drechsler, Ernst, Fischer, Glantz, Gräfenstein, Happ, Helfricht, Hengelhaupt, Hoepfel, Kehl, Keiner, Kind, Klett, Kneifel, Kolb, Kommer, König, Lämmermann, Langenhan, Lepper, Lesch, Luck, Mangler, Marr, Moritz, Muth, Paatz, Pickert, Reuss, Ritz, Sauerbrey, Schindler, Schneider, Schilling, Schlütter, Schramm, Schübler, Sontag, Spörer, Stockmar, Triebel, Weihrauch, Weiß, Will, Wirsing, Wolf, Zerr, Zink, Zöller, and dozens of smaller, lesser-known firms.

Erfurt also deserves mention, with the Königlich-Preussische Gewehrfabrik Erfurt and several commercial firms in the small arms business there from 1900-1945.[4] Lastly, the Thuringian town of Sömmerda was the ancestral home of the modern conglomerate Rheinmetall AG.

It is interesting to note that many of the larger firms involved in the manufacture of firearms also produced traditional consumer goods. J.P. Sauer for instance also made cutlery. Many of the gunmakers made bicycle components or even complete bicycles. Rheinmetall, Walther, Merz, and others made business machines and/or typewriters to keep themselves afloat before and between wars. Mauser not only made typewriters and adding machines, but for a short time even made taxi cabs. The proprietary logos and trademarks they used to brand and market firearms were often used on their other products as well.

Optics

Germany has long been the world's leader in the design and production of high quality optics, and many regions or towns within the country have claimed to have been the birthplace or "capital" of the industry.

Brandenburg certainly can lay claim to the title, with the establishment of Emil Busch AG, Rathenow in the latter half of the 19th century. An entire industry subsequently developed around Busch. Indeed, by 1896 there were 163 optical businesses registered in and around Rathenow, many of which developed out of small workshops of off-duty Busch employees. Some of the more prominent of these "cottage" firms include the Laro, Nitschke, Ruhnke, and Runge & Kaulfuss. F.O.G., Pecar, P. Köhler, Franz Talbot, Fritz Volk, R. Fuess, and Oigee also operated in the region, in and around the capital city of Berlin.

Other areas developed smaller but just as vibrant optics industries. The Erneman company operated for a time in Dresden, and of course Voigtländer & Sohn was established in Braunschweig.

Wetzlar and Kassel in Hesse and Jena in Thüringen (Thuringia) are perhaps the most famous of the group, and their histories have been tied to the production of optical glass, lenses, scopes and binoculars since at least the mid-1800s. E. Leitz and Hensoldt dominated the area in and around Wetzlar, but also spawned the Enuro, Hartmann, and Krombach firms. Indeed, Wetzlar began billing itself as the "Stadt der Optik" well before the turn of the previous century. Fullgrabe, Ruf, Dr. Wohler, Hertel & Reuss, Heine, and Spindler & Hoyer all sprang up in and around Kassel. By the eve of the Second World War, these:

"...firms engaged in the manufacture of the same item periodically consulted with each other and thereby exchanged ideas freely, methods of production, etc. Most of the smaller firms were subcontractors of the four or five larger firms and only made parts or sub-assemblies."[5]

Carl Zeiss founded his firm in Jena in 1846, which had by 1875 expanded to 60 workers. This eventually grew into the enormous Zeiss AG concern, which employed thousands and essentially controlled the entire German optical industry by the end of WWII:

"...Optical problems were referred to Zeiss, Leitz, Steinheil, etc. and the three above named did by far the most of all optical design for the German Army. Early in the war period such matters as which firm was most competent to manufacture a certain item was decided by an individual or group within the government, but gradually such questions of policy were either made by or greatly influenced by the Zeiss firm at Jena."[6]

Of note, over time most of the firms mentioned previously developed trademarks based on stylized optical prisms or lenses in one form or another.

Nordrhein Westfalen- Plastics

Thüringen (Thuringia), and to a much greater extent Nordrhein-Westfalen (Westphalia), for instance, and the town of Kierspe in particular, were centrally important to the production of phenolics, and the processing of plastics dominated the local economy. In fact, on the eve of WWII about half of the German factories and manufacturers of phenolic goods were located in or near the area. Kierspe alone contained thirty-six companies involved in processing phenolics, and its plastics industry was equipped with some 400-500 phenolic pressing machines. Plastics were increasingly important to all of the firms involved in manufacturing small arms, optics and edgeware, and a brief history of this industry, codes, and trademarks is discussed in the author's other Schiffer book *Kunststoffe: A Collector's Guide to German WWII Plastics and their Markings.*

Solingen- Cutlery

The area around Solingen, Germany, has been known for fine cutlery for centuries. Founded in 1374, the town fast became well known as a metal-working hub. This was due in large part to the abundance of high quality iron ore in the area, timber for charcoal, and flowing water, which were necessary to power large grindstones and hammer forges. The "Solingen Method" was established by guilds, which divided tasks within the area among specialized

Fig. 4: Skilled cutlery grinders in a small Solingen workshop ca. early Twentieth Century. Note the child apprentices in the group. *Courtesy the late William R. Smith.*

labor groups, such as sharpeners, polishers, forgers, and finishers. Soon Solingen was contracted to supply all of the knives and swords they produced to the region's richest trading center, Köln (Cologne). The quality of blades made in Solingen became so well known that the town name became an international brand name, and even today "Solingen" is a legally registered and protected trademark.

By the eighteenth century Solingen firms, such as Weyersburg and J.A. Henkels, were well-established and already using trademarks on cutlery quite similar to the ones in use today. During the same period, hundreds of other firms involved in the cutlery business were established. Steam-powered machines and forges were introduced in the 1850s, and companies such as WKC, Carl Eickhorn, C.W. Linder and Böker in Remscheid became even more revered for their fine-quality tableware, knives and sabers, many of which were used in the Franco-German War of 1870-1871.

During WWI, most of the large and smaller firms in the region produced cutlery, such as fighting knives and bayonets for the Kaiser's military. After the war, the industry and region struggled through the inter-war depression. Most of those that survived this period did so by producing tableware, razors, knives for hunters and other outdoorsmen, or edge ware for foreign military sales.

Then, domestic business simply exploded as the National Socialists gained prominence in Germany. The NSDAP and all of its subordinate organizations had an almost insatiable thirst for knives, swords, and ceremonial dress daggers. After they gained power in 1933, the Nazis gradually subjugated all governmental departments, from the police to the diplomatic service. Each in turn developed its own peculiar dress dagger or sword, most of which were made by firms in and around Solingen. The Hitler Youth, SA, SS, and growing Wehrmacht were the region's largest customers, and most of the cutlery makers prospered and expanded their facilities.

During WWII, many of the cutlery firms shifted their production from dress and ceremonial cutlery to fighting knives, bayonets, and other products for the expansive military. During this period the use of trademarks fell from favor, and the Solingen knife makers were ordered to adopt the German military letter code system. The area in and around the town was bombed extensively by the Allies, during which the historic center of the town was destroyed. Despite this, many of the firms produced cutlery for the war effort until the end of hostilities.

After WWII the larger, well-established companies quickly rebuilt and restructured. Many of the smaller firms disappeared or their assets, to include their trademarks, were acquired by others. As a result, many cutlery trademarks have changed owners several times. Of the hundreds of firms and trademarks known to be in circulation between 1900-1945, over 200 are listed or illustrated in the text.

3

Some General Notes on German Tradenames and Trademarks

Companies tended to develop their brands, tradenames, and trademarks over several years, decades or even centuries. Most of the early markings were simple and usually included something like the owner's initials, abbreviations, or acronyms of the company name and/or location. As time passed and markets changed and evolved, so too did the logos and trademarks.

Many trademarks evolved around some sort of a main figurative element, such as a banner, barrel, animal, geometric shape, et al, with the majority at least incorporating the name of the company and perhaps the location.

Companies frequently registered and used multiple trademarks and tradenames. In addition, a good number of larger companies often allowed smaller firms and sub-contractors to make goods under a certain name or trademark under license. Conversely, many smaller and lesser-known firms made goods that were sold to larger firms and then marketed under the latter's more established and familiar trademarks.

Certain common words or names have been claimed and used by many different companies over the years. The description "Original," and tradenames like "Geha," "Hubertus," "Tell," and "Venus" come to mind. Perhaps the most striking example of this is the name "Mars." Bergmann used the name to market pistols, and the Czech firm Posumavska Zbrojovka sold different pistol designs marketed with the same name. "Mars" was also used by Runge & Kaulfuss of Rathenow to sell binoculars, Venus Waffenwerk, Zella-Mehlis on air rifles, and Weyersberg, Kirschbaum & Cie (WKC) of Solingen to market tools and cutlery.

In any event, tradenames, trademarks and logos were used in a variety of ways beyond advertising. Most firms incorporated their proprietary markings as illustrations on their official stationary. Others marked their trademarks and logos on company property to discourage pilfering. And many small arms firms used one version or another of their trademarks as actual internal factory proof marks to indicate that a component or entire weapon was assembled and functioned to the established internal standards of the firm.[7]

4

Confusion Over Trademark Transfer and Ownership

Trademarks, and the companies they represent, can be valuable assets, and are subject to transfer and sale. Many of these have indeed changed hands over the years, and it is sometimes difficult to determine just what firm owned which trademarks during any given time.

Just one example is the cutlery industry in Solingen. Prior to the end of WWII there were hundreds of firms operating in the knife-making region. After the war some of the larger firms, such as Robert Herder, Henkels, and Eickhorn, quickly acquired many of the smaller firms or their assets, to include their proprietary names, trademarks and logos.

Another excellent example of how convoluted and confusing tradenames and trademarks can be is the Ludwig Loewe & Cie/DWM conglomerate. Loewe was primarily a financial holding company headquartered in Berlin that had ammunition, machine and tool making subsidiaries. In 1884, the arms-making firm of Gebr. Mauser & Cie of Oberndorf got into financial difficulty. The Württembergische central bank forced a reorganization of the company and appointed its own business manager. The founder, Paul Mauser, was left with only a minority interest in the new "Waffenfabrik Mauser AG," but still retained technical and creative control. The controlling interest in the company was soon passed on to Ludwig Loewe & Cie, Berlin. At this time, both companies used trademarks which consisted of rather conventional (for the time) stylized initials.

Loewe also had a small stake in the creation of the Belgian "Fabrique Nationale d´Armes de Guerre." "FN," as it was (and still is) known, was developed principally to manufacture small arms for the Belgian military. In 1896, following a number of disputes with FN, Loewe simply took control of the company by acquiring a majority of shares in the Belgian firm. This left Loewe as the "corporate parent" of Mauser, FN, the Hungarian FEG concern, and a number of ammunition makers and their respective names, logos and trademarks; but the marriage didn't last long. Following a scandal over bribing of Government inspectors to pass defective Gewehr 88s at Loewe's Berlin plant during the early 1890s (and resultant anti-Semitic accusations of treachery against the Loewe brothers personally) the firm decided to distance itself from ammunition and arms making.

Thus, in 1896 Loewe combined its arms manufacturing interests with the Rheinisch-Westfälische Pulverfabriken of Köln, Pulverfabrik Rottweil-Hamburg, and the Deutsche Metallpatronenfabrik of Karlsruhe to form DWM (Deutsche Waffen und Munitionsfabriken) as a separate organization. Loewe itself remained an independent company, still primarily involved with machine tools, and increasingly with electrical machinery during the early 20th century.

The new DWM ammunition and arms-making consortium originally used a trademark that consisted of the company initials written in cursive script. Eventually, though, DWM developed a flaming bomb logo for ammunition and another logo that consisted of the firm's initials within a figurative element representing a "Tonne" barrel, traditionally associated with storing and shipping finished gunpowder. The German Trade Mark Registry also described the DWM trademark as a "Tonne [barrel]" when it was officially registered. The barrel mark was used for both ammunition, firearms and other products.

Some of DWM's subsidiaries followed suit with this general corporate concept. Mauser changed its previous trademark to a now-familiar "Mauser within a stylized barrel." Later, the "Niccoro" brand of ammunition was developed by DWM, Karlsruhe (the former Deutsche Metallpatronenfabrik), which also made use of the brandname within a stylized gunpowder barrel.[8]

DWM later gained control of Waffenwerke Oberspree, Kornbusch & Company, Berlin (the early ancestor of the WWII "Spreewerk") in 1916. After WWI, in an attempt at survive the prospect of being broken up by the victorious Allies, the DWM tradename and logo were dropped, and the firm assumed the name "Berliner-Karlsruher Industrie-Werke A.G.," or "BKIW" for short. The consortium operated under this name from 1922-1936, after which it simply reverted back to its traditional and familiar "DWM" initials and trademark.

Berlin-Karlsruher Industrie-Werke

DWM

MAUSER

DWM

NICORRO

ROTTWEIL

Fig. 5: The evolution of the predominant trademarks and logos of subsidiaries that were controlled during the era by the Loewe firm and later DWM.

5

The Fall of Trademarking and the Rise of Military Number and Letter Codes

The use of tradenames and trademarks dominated German industry for the first three decades of the 20th Century. Then, in 1933, the infamous Adolf Hitler and the German National Socialists seized power and a short time later announced that Germany was disavowing the Treaty of Versailles and would begin to rearm. Soon, the German military-industrial machine was in full expansion, which did not go unnoticed by its potential adversaries.

Manufacturers of arms and ammunition were soon instructed to adopt and mark their products destined for the military and para-military organizations with letter/number codes instead of their traditional company trademarks and logos in an attempt to cloak their point of origin or manufacture.

In regard to small arms, the Simson firm in Suhl was the only company authorized by the Treaty of Versailles to make small arms for the German military. When the Nazis ordered clandestine production to commence at other factories they followed a general policy of secrecy. Cloaking the actual location by grouping them all under Simson's "S" designation, they also added an additional number suffix to designate the true point of origin. If a potential adversary saw a rifle marked "S/42" dated from 1934 onward, for instance, it would be explained that the rifle had come from Simson, as obviously indicated by the "S." "S/42," however, did not really represent Simson, but rather is widely known to have been used by the Mauser firm. S/27, S/147, S/237, and S/243 are other examples of well known rifle makers. The letter "P" is likewise known to have represented Polte Patronenfabrik of Magdeburg, the only company in Germany authorized under the treaty to produce ammunition, or "Patronen." Under this same secretive policy, other companies later tasked with cartridge production were also assigned the umbrella letter "P," but this was followed by their own numeric designation. Later, the letter prefixes were dropped from these clandestine numbers. Dropping the "S," small arms makers used numeric designators, such as "27," "147," and "237" for a short time in the late 1930s.

Optics, edgeware and other manufacturers, however, were not specifically mentioned in or restricted by the treaty. Most of these firms simply continued to use their established abbreviations, brand names, trademarks or logos on their products for both the military and domestic commercial and export markets.

German small arms and ammunition manufacturers were next ordered by law on July 8, 1939, to substitute an assigned number in place of their logos and trademarks, but most of the prominent firms had already been complying since the early days of the Third Reich. Then, on September 1, 1939, Hitler ordered the *Wehrmacht* to invade Poland. Western Europe was next, and although the nation was at war and most of German industry shifted their efforts from consumer goods to military items, industrial security within the country was still pretty *laissez faire*. In this early period of "phony" or "easy" war, most optics and edgeware were still being marked as they had been before, often with the firm's location, proprietary logo or trademark.

The British, though, first began small scale leaflet dropping and actual sporadic bombing of German targets in the autumn of 1939. RAF aircrews next conducted their first large-scale bombing of Berlin (in retaliation for an accidental bombing of London by disoriented Luftwaffe crewmembers) in August 1940, and by the end of the year had dropped some 14,000 tons of ordnance over the country.

These raids were concerning to the Nazi hierarchy, and many industrialists as well. It was apparent that the Allies were not just going to strike at military complexes, key infrastructure, and traditional heavy armaments manufacturers, but any industrial target of opportunity that might be important to the German war effort. Knife makers, foundries, precision optical manufacturers, and a myriad of others that had not previously needed shielding from the defunct Inter-Allied Military Control Commission were at this point susceptible to identification and attack.

Further complicating matters was the previously discussed fact that certain German industries had developed within specific regions over centuries, with many firms clustered in small areas. These locations were not difficult for Allied intelligence and targeting officers to identify through examination of pre-war advertising and individual items bearing a particular maker, logo, brand name or trademark.

As a result, a new military letter coding system, vastly broader than the older numerical method, was developed by the nation's armaments hierarchy in an attempt to better cloak the identity and location of key manufacturers. These military letter codes were implemented in November 1940, beginning with the group "aaa-aaz," and would eventually reach "ozz" and beyond. Most of the firms involved in small arms and optics production were also assigned a Reichs Betriebs Nummer (National Contractor Number) or "RB Number," which they tended to use for contracts, correspondence and stationary.[9] During this same timeframe, many manufacturers of field gear, leather, clothing and other items began marking their RB Number on their goods in lieu of military letter codes or their traditional names and trademarks.[10]

Compliance with the new systems varied considerably, especially during the early years of the war. The military letter code and RB systems did, however, essentially put an end to the widespread use of brand names and trademarks on products destined for the military between 1941-1945.[11][12]

Fig. 6: The progression of markings on the toggles of four Mauser P.08 pistols. Prior to 1940, manufacturers of military arms and equipment used their trademark logos and company names, and then were ordered to adopt the letter/number code system, followed by the number code system, and finally the military letter code system. The sequence is illustrated here, from the top to bottom: Mauser's stamped trademark is seen on the example produced in 1936, then the next year the "S/42" code, then the letter is dropped, leaving only "42" as a code, and finally, the "byf" military letter code. After 1940 manufacturers were supposed to use their assigned military letter code in lieu of their logos and trademarks on products destined for the military, but compliance with this directive varied from company to company. *Courtesy Dr. Geoffrey Sturgess.*

Fig. 7: 1930-1940s era *Reinigungsgerät* 34, or "Cleaning Equipment 34" cleaning kits, also serve to illustrate the transition from names/trademarks to numeric codes and finally to military letter codes.

Fig. 8: In this instance, the firm of Gustav Appel Maschinenfabrik, Berlin-Spandau produced RG34s marked with the number code "64," then later went back to simply stamping the company name "G.Appel" during WWII, and finally made the transition to its "cnx" military letter code.

6

Notes on Post-WWII German Tradenames and Trademarks

The decades following the end of the Second World War saw the establishment of two separate German states: the Federal Republic of Germany (West Germany) and the German Democratic Republic (the DDR, or East Germany). The Western Zone was occupied and controlled by the victorious Allies, while the Eastern Zone fell under Soviet occupation.

The Western Allies gradually allowed the reestablishment of German self-government in the Western Zone and loosened their control over the country's economy and industry. Many prominent firms were back in business, again using their same pre-WWII trademarks and logos. In 1954-1955 there appears to have been an absolute flurry of registration and re-registration of pre-1945 trademarks as the West German economy recovered.

Firms still located in the East, however, were not as fortunate. Nearly all of the small arms companies, for instance in Thüringen, were dissolved, and their machinery shipped to the USSR as reparations. Many of these companies' key personnel fled westward as a result and reestablished their firms in new locations. What remained in the Soviet Zone by way of equipment and real estate was nationalized by the East German communists into industrial collectives, called Volkseigener Betrieben, or "People's Enterprise" (abbreviated "VEB").

What this all amounted to was a number of companies and collectives with the same or close to the same name on both sides of the Iron Curtain. These firms shared the same lineage prior to the end of WWII, and each could and did rightly lay claim to all of the pre-war tradenames and trademarks of the legacy firm. The Walther and Sauer companies are good examples of this post-WWII confusion, and it took many decades of legal wrangling to sort most of it out. The battle between the two Zeiss firms was particulary bitter. In some cases the issues weren't settled until the fall of communism in 1989.

33 **650432** **S 2565**

Fig. 9: Firms used the word "Original" to differentiate their products from "knock-offs" manufactured in China, Spain, and other countries. When Germany capitulated and was divided after WWII, many executives and workers from companies that then fell under Soviet occupation fled westward and started their companies anew in the Western Zone. For a time there were companies in business on both sides of the Iron Curtain laying claim to and using the same pre-1945 trademarks. This post-WWII, revamped and re-registered trademark of J.P. Sauer also incorporates the word to differentiate the company from the remnants of its former facilities in the communist DDR.

7

Names and Tradenames of Selected Small Arms, Optics, and Edgeware Manufacturers

The following section lists companies, abbreviations, and tradenames compiled during the author's research in alphabetical order. Tradenames and abbreviations are listed in upper or lower case depending on how they were observed as written, shown, or used between 1900-1945. Some foreign firms that supplied large numbers of small arms to or under the supervision of the German military and post-WWII firms and tradenames have been included for collector interest and clarity. A brief description of each company, corporate ties, predominant end-products, and additional information are also provided, if known at the time of this writing.

*: "asterisk"; not a trademark or logo. An eight-armed asterisk was used by numerous WWII-era small arms firms to denote a part or component that was initially made out of specification but was deemed repairable or reworkable. The asterisk marks are commonly found on K98s and G43s made by Gustloff, other K98 manufacturers, and P38 pistol makers. The asterisk may be confused with 6-armed star logo used by Gebr. Böhler & Co. AG, Edelstahlwerke to mark its products.

Aba: "Aba-Metall," Allgemeine Berliner Automaten- u. Metallwarenfabrik GmbH, Berlin; a manufacturer of small arms components and .22 caliber training kits for the P08 and other goods during the National Socialist era.

Abesser & Merkel: Abesser & Merkel Gewehrfabrik, Suhl, maker of sporting arms in between about the 1920-1940s. The company was founded and owned by Paul Abesser and Ernst Merkel. After WWII the firm's assets were consolidated into the GDR's "V.E.B. Gebr. Merkel, Suhl" (see also *Merkel*).

Adamy: Adamy Gebr., Suhl, manufacturer/retailer of sporting arms, founded in 1921.

Adler: Trade name for pistols made by Adler Waffenwerke, Zella St. Blasii, circa 1900-1910.

ADLER FLINTE: Tradename used by Anton Ferdinand Korinski, Erlangen to market high quality Drillings and sporting arms from 1883-1920.

Fig. 10: The markings and trademark on a fine example of an Adler Pistol made by Adler Waffenwerke, Engelbrecht & Wolff, Zella St. Blasii (later, Zella-Mehlis). *Courtesy Dr. Geoffrey Sturgess.*

Adler Waffenwerke: Adler Waffenwerke, Engelbrecht & Wolff, Zella St. Blasii; a subsidiary of Adler Fahrradwerke, a manufacturer/retailer of bicycles and small arms, circa 1905-1910.

AEG: Abbreviation for Allgemeine Elektrizitäts-Gesellschaft. See *Allgemeine Elektrizitäts-Gesellschaft*.

Aesculap: Aesculap Aktiengesellschaft für Feinmechanik, vormals Jetter & Scheerer, Tuttlingen; primarily known for precision surgical instruments, Aesculap also made cutlery and dress daggers during the National Socialist era. The firm survived WWII and continued to make surgical instruments and other products well after the war.

AGFA: AGFA Kamerawerk (or "Actien Gesellschaft Für Anilin-Fabrikation") was founded in 1867. By 1898, AGFA had become deeply involved in manufacturing plates and films for X-ray research. AGFA was acquired by Bayer, which had itself become a subsidiary of I.G. Farben. In 1936 AGFA introduced the "Agfacolor" color photography process. During WWII the firm concentrated on camera and film production, but also manufactured thousands of ZF4 rifle telescopes marked "bzz" at its Munich facility.[13]

Fig. 11: AGFA's trademark as it appeared in 1940s print advertising.

AJACK: Tradename used to market rifle telescopes made by Adolph Jackenkroll Optische Anstalt, Berlin, 1930-1940s.

Akah/AKAH: Tradename/abbreviation used by Albrecht Kind, Berlin-Nürnberg; firearms and ammunition retailer and maker of holsters, scope cases and leather goods. See also *Kind, Albrecht*.

Alcoso: "Alcoso" GmbH, Alexander Coppel, Stahlwaren- u. Waffenfabrik, Solingen; cutlery manufacturer (daggers, knives, bayonets, et. al.) that used trademarks and logos of many variations, such as the "Coppel" tradename and "ACS" abbreviations. Established in 1821, the firm was named after the prominent Solinger, Gustav Coppel, who was Jewish. Control of Alcoso remained in family hands until the 1930s, when Nazi pressure forced Alexander Coppel to sell the company. Sophie Coppel, Alexander's sister, fled to Switzerland in 1934. Carl Gustav Coppel committed suicide in 1941, and his daughter Anna died in the Ravensbruck Concentration Camp. Alexander Coppel was deported to the Theresienstadt (Terezin) Ghetto in 1942 and died there just weeks later. After the war, the surviving members of the family regained control of the company, which was still in business at the time of this writing.

Fig. 12: The Alcoso trademark and tradename of Alexander Coppel, Stahlwaren u. Waffenfabrik, Solingen, frequently seen on WWII-era cutlery. *Courtesy Robert Jensen.*

ALFA: Trade name used by Adolf Frank, Hamburg, an arms distributor-exporter-importer in operation around 1900-1915.

Fig. 13: Detail from a WWI-era P08 "snail drum" magazine made by Allgemeine Elektrizitäts-Gesellschaft, or "AEG." *Courtesy Dr. Geoffrey Sturgess.*

Allgemeine Elektrizitäts-Gesellschaft: Allgemeine Elektrizitäts-Gesellschaft; a huge manufacturer of electrical appliances and other goods and services. AEG was founded in 1883 by Emil Rathenau in Berlin as the Deutsche Edison Gesellschaft, or "DEG." The company prospered making appliances and struck a deal to supply a large portion of the city with electricity. In 1894 the name was changed to Allgemeine Elektrizitäts-Gesellschaft, better known simply as "AEG." The company quickly became a huge conglomerate with subdivisions involved in the automotive, aircraft, plastics and electrical industries. In the context of this text, AEG is known to have made small arms components and particularly 08 "snail drum" magazines marked with the firm's hexagonal trademark during WWI. During WWII the firm was heavily involved in all sectors of war production, with numerous facilities in and around Berlin, as well as Annaberg, Henningsdorf, Nürnberg, Stuttgart, Litzmannstadt, Wildau, Wien, and others inside and outside Germany.

Altenbach: Peter Altenbach, Solingen; cutlery and dress dagger manufacturer during the Nazi era.

Altenbach: Peter Altenbach und Söhne, Schanenwerk, Solingen-Ohligs; maker of cutlery and daggers during the Nazi era.

AMBERG: Königliche Gewehrfabrik Amberg, Bayern; an Imperial Bavarian, state-owned small arms production facility/arsenal first established in 1801. The arsenal made cutlery, Gewehr 71 and 71/84s, Gewehr 88s, Gewehr 98s, and the Kar98a, and produced bayonets made from blade blanks supplied by F. Herder in Solingen prior to the end of WWI. Between 1918-1920, the directors of the facility tried to keep the workforce employed by producing commercial hunting arms and other goods for the civilian market (many of these marked "Gwf. A"). The arsenal was then essentially dismantled due to the restrictions in the Treaty of Versailles, and much of what remained was absorbed into the Deutsche-Werke. See also *Deutsche-Werke*.[14]

Fig. 14: WWI Gewehr 98 as manufactured by the Amberg arsenal. *Courtesy Robert Jensen.*

Anker: Anker-Werke AG, Bielefeld; a manufacturer of business and sewing machines founded by Carl Schmidt in 1876 as the "Bielefelder Nähmaschinenfabrik Carl Schmidt." Schmidt soon branched out into bicycles and adding machines, and the company name was changed to the "Anker-Werke AG" in 1906. Anker made bayonets during the Imperial period among other products, and employed about 1000 workers in the 1920s and over 2000 by 1938. During WWII the firm made small arms components, was a principal manufacturer of rifle grenade launchers (Gewehrgranatgerät), and submitted a prototype self-loading rifle, the "G43(A)."[15]

Fig. 15: Illustration from an Anker-Werke prospectus which incorporated the firm's longtime "anchor" trademark.

Anschütz: Bruno Anschütz, Zella-Mehlis; made sporting arms in the 1920s.

Fig. 16: The J.G. Anschütz trademark as seen on this 4 mm Wehrsportgewehr. *Courtesy Dr. Geoffrey Sturgess.*

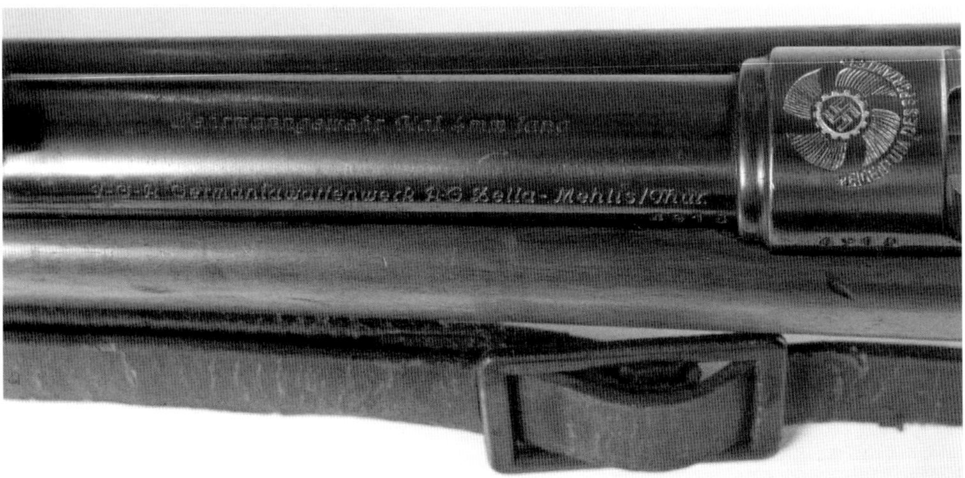

Fig 17: A "Wehrmanngewehr" 4 mm caliber training rifle made by J.G. Anschütz, Germaniawaffenfabrik, Zella-Mehlis. In this case, the firm did not use its familiar "JGA" trademark on the receiver, but rather, engraved the National Socialist logo and "Eigentum des Sportamtes," or "Property of the Sport Office."

Anschütz: J.G. Anschütz, Germania Waffenfabrik, Zella-Mehlis; founded in 1856 by a gunsmith's son, Julius Gottfried Anschütz, the company originally manufactured Flobert and pocket pistols, "Teschings," flare pistols and shotguns. By the eve of WWI the company had over 200 employees. After the war the firm became famous as a manufacturer of flare pistols, small arms components and small caliber training and target rifles. During WWII, nearly all of the company's production capacity was tied up in the manufacture of MP44 components.[16] The firm was reestablished after the war in Ulm, West Germany.

Anschütz: Udo Anschütz, Zella-Mehlis; manufactured sporting arms and "Free-Pistolen" in the 1920-1930s.

Appel: Gustav Appel Maschinenfabrik, Berlin-Spandau, maker of tools, grenade launchers, carriers, containers, and other stamped metal goods. In addition, Appel was the patent holder and principal manufacturer of the RG34 small arms cleaning kits. Appel is also known to have finished machine gun barrels and submitted a single-shot, sliding-block rifle proposal for arming the Volkssturm, 1944.

Argenta: Argenta-Gesellschaft, Solingen; cutlery and dagger maker during the Nazi era.

Arbeit: Carl Arbeit, Hamburg; a major exporter and distributor of small arms.

ARGUS: Tradename used to market rifle telescopes made by Karl Kahles, Optiker, Wein.

ARMESO: Tradename used by Arthur Melcher, Solingen-Merscheid to market daggers and other cutlery.

ARMINIUS: Tradename used in marketing revolvers made by Friedrich Pickert, Zella-Mehlis prior to WWI. After WWII this tradename was used by Herman Weihrauch Sportwaffenfabrik, West Germany (the post-war West German incarnation of Herman Weihrauch Gewehr und Fahrradteile Fabrik, Zella-Mehlis, Thüringen). See also *Weihrauch*.

Aschenbach: Aschenbach; a maker of cutlery during the Nazi era, frequently marketed under the "Ankermast" tradename.

Askania: "Askania"-Werke AG, Berlin-Friedenau, maker of instruments, gauges, range finders and other precision optics. Askania was founded in 1921 by Carl Bamberg, a former employee of Zeiss AG.

Fig. 18: The WWII-era letterhead of the Askania-Werke AG, Berlin-Friedenau.

ASS: Abbreviation of August Schüler, Suhl. See *Schüler*.

Assmann: Assmann & Söhne, Lüdenscheid; a prolific manufacturer and distributor of medals, awards accoutrements and other products for the para- and military during WWII.

ASSO: Tradename used by Arthur Schüttelhöfer & Co., Solingen to market cutlery during the National Socialist era.

Astora: Tradename used by Gustloff-Werke, Suhl to market sporting arms in the late 1930s. See also *BSW* and *Gustloff-Werke*.

ASTORIA: Tradename used on sporting arms, bicycles and other products made by Simson & Co, Suhl and later Berlin-Suhler Waffen u. Fahrzeugwerke/Gustloff-Werke, Suhl. See also *BSW* and *Simson*.

ASTRA: Tradename of Astra-Unceta y Cia. S.A., Eibar (later Guernica), Spain. The company was formed under the name of Pedro Unceta y Juan Esperanza in 1908 in Eibar, and only later moved to Guernica (ca. 1913). In about 1926 the business was renamed Esperanza y Unceta and was changed again to Unceta y Cia in ca. 1926. Thousands of Astra pistols of various types were purchased and delivered to the German military during WWII.

Fig. 19: The Astra trademark as roll-stamped on this WWII German-contract Model 600 pistol. *Courtesy Dr. Geoffrey Sturgess.*

Astrawerke: Astrawerke, Chemnitz; a major supplier of various milled and stamped components for the K98 rifle during WWII. The firm was founded in 1921 by John Greve and originally made business machines. Eventually it was pressed into war production and made K98 receivers and components, as well as parts for the MP44 and late-war Volksgewehre. After WWII the Astrawerke was merged with the remains of the Wanderer firm into the "VEB Mechanik Astrawerke" and resumed production of adding machines, calculators and eventually computers. The company was liquidated after German reunification in 1990.

Auer: Auer-Gesellschaft AG, Berlin; a firm involved primarily in the manufacture of gas masks and stamped sheet metal canisters, but also made numerous sheet metal products for the small arms industry. Auer had subsidiaries in Guben and Danzig and survived WWII.

AWS: Abbreviation used by Alexander Coppel, Stahlwaren- u. Waffenfabrik, Solingen, maker of cutlery and K98 bayonets. See also *Alcoso*.

A.W.Z.: Abbreviation for Albin Wahl, Zella-Mehlis. Wahl manufactured a conventional 6.35mm pistol with the "Stern" trademark and the abbreviation "A.W.Z." on the grips around the 1920-1930s.

Axt und Haverfabrik: Axt und Haverfabrik, Solingen; cutlery manufacturer during the Nazi era.

Aydt: Refers to a rifle system used to make ornate high-quality Schützen rifles around the turn of the 19th/20th century. Carl Wilhelm Aydt originally developed and patented the system in 1881. He further modified and improved it over the years while working for C.G. Haenel, Suhl. See also *Haenel*.

Backhaus: F.W. Backhaus, Solingen; cutlery and dagger manufacturer during the National Socialist era.

Bahel: Walter Bahel; cutlery maker during the Nazi era.

Bahrl: Julius Bahrl, Solingen; cutlery and dagger manufacturer during the National Socialist era.

Bahrl: Fa. Walter Bahrl, Solingen; cutlery maker during the Nazi era.

Balke: Rich. Balke & Söhne, Solingen; cutlery and dagger maker during the National Socialist period.

BAR PISTOLE: Tradename used by J.P Sauer & Sohn, Suhl to market an interesting four-shot pistol designed by Burkhard Behr and marketed ca. 1898-1910.

Barbier: Etablissement Barbier, Benard et Turenne, Paris; maker of precision optics and rifle telescopes under German occupation, WWII.

Barella: H. Barella, Berlin; supplier of sporting/hunting arms, many of which were exported abroad.

Barth: Carl Barth, Radeberg; a principal supplier of K98 rifle components and other military goods during WWII.

Bassat: Albert & Heinrich Bassat, Solingen; cutlery maker during the Nazi era.

Bartels: Fr. Jacob Bartels, Wiesbaden; a distributor and retailer of small arms.

Barthelmess: Fritz Barthelmess, Muggendorf; cutlery and dagger manufacturer during the National Socialist period. The firm used the "BAVARIA" tradename.

Bastian: Gunter Bastian; an edged weapons maker during the National Socialist era.

Bauer: Karl Bauer & Co., Suhl, a retailer and distributor of firearms, many of which were made elsewhere (such as August Menz) and simply carried the "Kaba Spezial" brandname.

Bauermann: Bauermann & Söhne, Hilden/Rhein, bei Düsseldorf; cutlery maker during the Nazi era.

BAVARIA: Tradename used by Fritz Barthelmuss to market edged weapons during the Nazi era.

Bayard: Pistol design/tradename manufactured by Anciens Etablissements Pieper, Herstal, Belgium, from 1908 until the later 1930s; a number of small Pieper-Bayard 1908 pistols were pressed into German service during the WWI period and bear Imperial proof marks.

Bayerl: Ludwig Bayerl, Büchsenmacher, Regensberg; manufacturer/catalogue retailer of Drillings and other sporting arms, circa early 1900s.

Becker: Eduard Becker, Kolumbuswerk, Solingen; cutlery maker during the Nazi era.

Becker: Gebr. Becker, Solingen; cutlery and dagger maker during the Nazi and post-WWII periods.

Becker: J.H. Becker, Solingen; a cutlery manufacturer during the Imperial era, known for Sg 98/05 bayonets.

Becker und Holländer: Becker und Holländer, Suhl; originally known as Becker & Holländer Waffenbau Suhl, was a manufacturer of pistols marketed under the trade name "BEHOLLA," derived by shortening both of the company names. Becker und Holländer started making the pistol in 1915 and manufactured approximately 45,000, most of which went to the German military in WWI.

Beckmann: Beckmann GmbH, Solingen; cutlery maker during the Nazi era.

BEHOLLA: Tradname for pistols made and marketed by Becker and Holländer, Suhl. Other companies manufactured the same design under licensing agreements as well. These included Stenda Werke Waffenfabrik, Suhl (marketed as the "STENDA"), H.M. Gering & Cie, Suhl ("Leonhardt") and Waffenfabrik August Menz, Suhl (marketed as the "MENTA").

Bell: Gebr. Bell, Solingen, cutlery and dagger manufacturer during the pre- and WWII era.

Ben Hur: Tradename used by Friedrich Herkenrath, Merscheid to market edged weapons during the Nazi era.

Bender: Carl Bender, Solingen; cutlery maker during the National Socialist regime.

Benz: G. Benz u. Sohn, Kiel; a distributor and retailer of small arms.

Beretta: Fabbrica d'Armi Pietro Beretta, Gardone VT (Brescia), Italy; a centuries-old arms maker. The company was founded in 1526, and has remained in family control ever since. By 1900 Beretta was making sporting and military arms and supplied the Italian monarchy with weapons during WWI. The firm survived the war and was well positioned when Mussolini came to power, eventually supplying his fascist military with SMGs, rifles and pistols. After the 1943 Italian capitulation, Beretta continued to make small arms for the RSI (Socialist Republic of Italy) and the Wehrmacht under German occupation.

Berg: Berg & Co., Solingen-Ohligs, maker of cutlery and K98 bayonets.

Berga: Seidenweberei Berga, C.W. Crous & Co., Berga/Elster; maker of precision optics and rifle telescopes during WWII.

Berger: Max Burger, Höxter; a manufacturer or distributor of dress daggers during the Nazi era.

Bergfeld: Edmund Bergfeld & Sohn, Solingen; manufacturer of edged weapons during the Nazi era.

Bergmann: Theodor Bergmann, Gaggenau, designed a number of self loading pistols and marketed pistols made by other firms, such as V. Christoph Schilling.

Bergmann: Theodor Bergmann Maschinen u. Waffenfabrik, Suhl and Gaggenau; small arms manufacturer. Founded by Theodor Bergmann, the firm made pistols and machineguns from 1892 onward (and sub-contracted the manufacture of others). Bergmann introduced a new pistol design nearly every year it was in business and used tradenames such as "MARS" and "Simplex," and usually marked its products with a nude "miner with lamp"

Fig. 20: An original Bergmann carton for the Simplex pistol with embossed tradenames. *Courtesy Dr. Geoffrey Sturgess.*

logo. Bergmann is often described as an early small arms genius, but contemporary sources indicate that he was more of an entrepreneur, with designers Louis and his son Hugo Schmeisser the actual talent in the company. Nevertheless, Bergmann and the Schmeissers worked together in developing the MP18/I for the German military, production of which began in 1918. In 1921-1922 the firm was acquired by the Aktiengesellschaft Lignose, which itself was traditionally involved in manufacturing gunpowder. After WWI Hugo and his brother Hans departed Bergmann and formed their own company, Industriewerk Aufhammer Koch & Co., Suhl. Hugo Schmeisser subsequently developed a number of SMG and pistol designs and protected them with patents under his own name. The Bergman firm survived the turbulent decades between world wars, eventually reorganized as Bergmann & Co. KG, Waffen-u. Munitionsfabrik. Near the end of WWII Bergmann & Co. KG was involved in the development of single-shot Volksgewehr prototypes. Bergmann's name survived the war, but the company switched from making small arms to producing plastic elastomers. At the time of this writing it was known as Theodor Bergmann Kunststoffwerk GmbH, Gaggenau. See also *Schmeisser* and *Lignose*.

Bergsneller: Bergsneller & Söhne; manufacturer of Drillings and sporting weapons.

Berlin-Karlsruher Industrie-Werke: Berlin-Karlsruher Industrie-Werke; the renamed post-WWI incarnation of *DWM*. This title was used from approximately 1922-1936. See also *DWM*.

Fig. 21: An early-1920s carton for a DWM M1923 pocket pistol displaying DWM's "new" name, Berlin-Karlsruher Industrie-Werke. The company would subsequently change back to DWM in 1936. *Courtesy Dr. Geoffrey Sturgess.*

Berlin-Lübecker Maschinenfabrik: Berlin-Lübecker Maschinenfabrik, Lübeck; one of five companies owned and directed by the industrialist, Bernhard Berghaus. The firm began manufacturing small arms in 1934. In 1935 the Heereswaffenamt awarded a contract to BLM to manufacture the K98k. BLM spent the remainder of 1935 tooling up and commenced manufacturing the K98k in 1936. Manufacture of the G41 rifle subsequently began in late 1942. By 1943 BLM was manufacturing torpedo components, G41s, LP42 flare pistols and other small arms. After finishing up their remaining G41 order in early 1944, BLM made the transition to G43 rifle production. The company only rarely used its "BLM" within a six-sided geometric figurative element on products it produced.

Fig. 22: A poor-quality image of a barrel straightening machine obtained from a CIOS report that nevertheless illustrates the rarely seen trademark of the Berlin-Lübecker Maschinenfabrik, Lübeck. *Courtesy US National Archives.*

Fig. 23: "BSW" stamped onto the receiver of an early Karabiner 98k. The firm was the only manufacturer to use its logo on K98k rifles well into the National Socialist era. *Courtesy Robert Jensen.*

Berlin-Suhler Waffen- und Fahrzeugwerke: Berlin-Suhler Waffen- und Fauhrzeugwerke GmbH (BSW), with offices in Berlin and Suhl; Small arms maker, aka "Simson & Co., Suhl." The "BSW" abbreviation and name was used by the Simsons before the company was officially seized by the National Socialists in November 1935. BSW made machineguns, rifles, pistols, and other small arms as well as bicycles and baby carriages. BSW was later incorporated into the Gustloff-Werke foundation. See also *Simson* and *Gustloff*.

Fig. 24: The same BSW logo stamped into a MG13 magazine manufactured just before WWII. BSW was eventually consolidated into the Gustloff-Werke. The eagle/4 marking is a Heer acceptance stamp.

Händler-Nettopreisliste 1939
für BSW-Fahrräder

BERLIN-SUHLER WAFFEN- UND FAHRZEUGWERKE

BSW

STIFTUNGSBETRIEB DER

WILHELM-GUSTLOFF-STIFTUNG

WAFFENSTADT SUHL

Waffenstadt Suhl: Anruf 2701, 2702, 2703, 2704, Drahtanschrift: Beswewaffen Suhl,
Reichsbank-Girokonto, Suhl 81 Postscheckkonto Erfurt 990

Fig 25. BSW's 1939 bicycle price list. Note that the company, which was seized from the Jewish Simson brothers, was absorbed by the Gustloff-"Foundation." *Courtesy Paul Shomper.*

Bern: Waffenfabrik Bern, tradename of Eidgenössissche Waffenfabrik, the Swiss State military manufacturer in Bern, Switzerland. Waffenfabrik Bern identified its products with a "W+F" marking.

Berns: Gebr. Berns, Solingen; cutlery and dagger maker during the National Socialist regime.

Berns: Hugo Berns, Solingen-Ohligs; cutlery and dagger maker during the National Socialist regime. The Berns firm sold cutlery under the "Hubeo" tradename.

Bertram: C. Reinhard Bertram & Sohn, Solingen; a cutlery maker first established in 1845 as the Solinger Stahlwarenfabrik. Bertram made razors, knives and daggers through the National Socialist era to the modern day.

Beuttenmüller: C. Beuttenmüller & Cie., Metallwarenfabrik, Bretten/Baden; a company specializing in stamped metal products. During WWII Beuttenmüller made containers, muzzle covers, and magazines for the MKb42(H) and MKb42(W) and other products for the military.

Bickel: August Bickel, Steinbach-Hallenberg; a cutlery and dagger maker during the National Socialist era.

Bijou: Tradename used to market small-caliber pistols made by August Menz, Suhl. See also *Menz*.

Billeter & Klunz: Billeter & Klunz, Werkzeugmaschinenfabrik und Eisengießerei, Aschersleben; a machine and tool making concern founded in 1857 by the company's namesakes Heinrich Billeter and Wilhelm Klunz. The firm employed 150 workers by ca. 1900, was involved in producing various metal wares for the military and commercial products in peacetime. During WWII, the firm employed some 1,200 machinists and metal workers and was a well known maker of anti-tank guns, K43 magazines, and made a small number of VG1 Volksgewehre in 1945. After the war it was reformed by the DDR authorities into the "VEB Werkzeugmaschinen Fabrik Aschersleben (WEMA)."

Bing: Gebr. Bing, Nürnberg; a manufacturer of stamped metal goods to include the tin-plate toys, and the P08 Trommelmagazin, or "snail drum magazine" ca. WWI. Products made by the Bing Brothers firm were often marked "B/N."

B/N: See *Bing* above.

Fig. 26: Detail from a WWI-era P08 "snail drum" magazine made by Gebr. Bing, Nürnberg. *Courtesy Dr. Geoffrey Sturgess.*

Bischoff: W. Bischoff, Braunschweig; a maker of precision optics to include rifle scopes. Bischoff was founded in 1903 and is thought to have made scopes prior to and probably during WWII, and was certainly in business marketing them after the war, but the company's history between ca. 1930-1945 is worthy of further research. Bischoff is also known to have made a wide range of scopes marketed under the names "DINOX."

BISMARCK: Tradename used by August Müller, Solingen to market cutlery during the Nazi era.

Bittner: Gustav Bittner, Waffenfabrik, Weipert/Erzgebirge (Vejprty); a Bohemian small arms maker, best known for making small arms components, early repeating pistols and Luftwaffe flare pistols.

Fig. 27: Detail from a Luftwaffe Modell-L double-barrel flare pistol, made by and marked with the Gustav Bittner trademark. The stylized eagle/109 is a Luftwaffe inspector marking. *Courtesy Dr. Geoffrey Sturgess.*

BKIW: Abbreviation for Berlin-Karlsruher Industrie-Werke; see same and *DWM.*

BLM: Abbreviation for Berlin-Lübecker Maschinenfabrik, Lübeck. See *Berlin-Lübecker Maschinenfabrik.*

Blohm & Voss: Blohm & Voss, Schiffswerft u. Maschinenfabrik, Hamburg. Founded in 1877 by Hermann Blohm and Ernst Voss, the company operated huge shipyards and had an aviation branch famous for its flying boats and the odd offset-fuselage BV 141 aircraft. By late 1944, however, little was left of the German surface fleet, and many shipyards were either sitting idle or in ruins. Thus, the company was ordered to make a version of the MP3008 submachine gun (based on the British Sten) for the Volkssturm. The firm used a trademark that consisted of the company initials within two circles.

Fig. 28: Blohm & Voss is best known for ship building, but also made a number of MP3008s near the end of WWII marked with the company trademark.

BOCH-DRILLING: Tradename used by S. Gunther, Suhl to market sporting arms prior to WWII.

Bochem: Ernst Bochem, Solingen; manufacturer of edged weapons during the National Socialist period.

Bochumer: Bochumer Verein für Gußstahlfabrikation AG, Bochum; a manufacturer of high quality steel for the aircraft, heavy gun and small industries during the National Socialist era.

Bock: Otto Bock, Hofl. Berlin; made rifle telescopes for the German military circa WWI.

Bodenstein: Julius Bodenstein, Steinbach; a cutlery and dagger maker during the National Socialist era.

Boehme: Dr.-Ing. Boehme & Co., Werksleitung Lüdenscheid, Lüdenscheid; a manufacturer of stamped metal goods, such as MG belt fillers, tools and K98 components.

Boger: Dr. M. F. W. Boger, Hamburg; an optical firm involved in WWII optical production.

Bö: Abbreviation/marking used on barrel blanks and other products made by Böhler & Co. AG, Edelstahlwerke. See also *Böhler*.

Boehme: Dr.-Inh. Boehme & Co., Werksleitung Lüdenscheid, Werk Minden; a bearing manufacturer that also made various military goods such as machinegun loading tools and K98 components.

GEBR. BÖHLER & CO.
AKTIENGESELLSCHAFT
EDELSTAHLWERKE

Fig. 29: The letterhead of the Gebr. Böhler & Co. AG, Edelstahlwerke, a maker of high quality steel for the small arms and other industries. Böhler frequently marked its products with a 6-pointed star/asterisk, which is also seen above. *Courtesy Jon Speed.*

Böhler: Gebr. Böhler & Co. AG, Edelstahlwerke, made high-quality barrel blanks made of "Böhler" steel throughout the WWI-WWII era. Böhler & Co used at least a dozen different trademarks and logos and had subsidiaries in Berlin, Bruckbacherhütte bei Waidhofen/Ybbs, Waidhofen, Düsseldorf, and Kapfenberg. Böhler used a number of marks over the years, to include its familiar six-pointed star/asterisk, the word " BÖHLERSTAHL," and/or the abbreviation "Bö."

BÖHLERSTAHL: a proprietary tradename of Böhler & Co. AG, Edelstahlwerke, often stamped on barrel blanks made of high-quality "Böhler" steel. Böhler & Co had subsidiaries in Berlin, Bruckbacherhütte bei Waidhofen/Ybbs, Waidhofen, Düsseldorf and Kapfenberg. See also *Böhler*.

Böhme: Gebrüder Böhme Nachf., Brotterode; a cutlery and dagger maker during the National Socialist era.

Böhmenachel: Gebr. Böhmenachel; an edged weapons maker during the National Socialist era.

Böhmische Waffenfabrik: Böhmische Waffenfabrik AG, Brno. In 1939, the Germans annexed Czechoslovakia and the Česká Zbrojovka (CZ) conglomerate, a longtime manufacturer of rifles, pistols and other small arms, fell under German occupation. CZ's name was changed to Böhmische Waffenfabrik AG and operated under German supervision until 1945 (though the "CZ" trademark remained in use). After the war, the firm was reorganized as a "people's enterprise" and continued operations under Czech communist control. CZ was again privatized after the end of the Cold War and is still active in the small arms market today. See also *CZ*.

Böker: Heinrich Böker, Baumwerk, Solingen; founded in Remscheid in 1829 by Hermann and Robert Böker, the company quickly became known for quality sabers. Hermann Böker later emigrated to the United States and founded Hermann Boker & Co. in New York. In 1865, Robert established companies in Canada and Mexico. Back in Germany, Heinrich Böker moved the original company a short distance away to Solingen and renamed the firm Heinr. Böker and concentrated on the production of razor blades, scissors and pocket knives. The company "tree brand" trademark was based on a large chestnut tree located close to the factory. Heinrich Böker first registered this trademark in 1869. The pocket knife business boomed around the turn of the twentieth century, with most of the production exported to North America. Böker expanded considerably in the following decades, making and subcontracting the manufacture of knives, utensils and dress daggers. Böker cutlery was especially valued by the National Socialists and the company prospered during the era. The Solingen facility, however, was completely destroyed during WWII and the family temporarily lost control of its trademark and tradename. The Böker facility was eventually rebuilt, and Böker remains in business at the time of this writing.

Böker: Karl Böker, Solingen: another member of the Böker family involved in the cutlery trade. Karl Böker was also a knife and dagger manufacturer during the National Socialist era.

Bolte & Anschütz: Bolte & Anschütz Jagdwaffenfabrik, Zella-Mehlis; manufacturer and retailer of sporting arms, components, sub-caliber inserts, and small caliber training rifles. Owned by Fritz Reuss from about 1900 onward, the company thrived during the 1920-1930s, but is not known to have survived beyond WWII.

Fig. 30: Period print advertisement of the cutlery firm Böntgen & Sabin, Solingen. Note that the firm had at least two recognizable trademarks. *Courtesy Paul Shomper.*

Böntgen & Sabin: Böntgen & Sabin, Solingen; maker of Sg 84/98 bayonets, ceremonial daggers and other cutlery during the Imperial-Nazi eras.

Born: Gebrüder Born, Solingen; a cutlery and dagger maker during the Nazi era.

Borsig: August Borsig, Berlin-Tegel. See *Rheinmetall.*

Fig. 31: The trademark of August-Borsig, Berlin-Tegal, a maker of locomotives, after the firm was acquired by Rheinmetall AG in 1933. The companies became known as Rheinmetall-Borsig after the merger. These markings were observed on a 10.5 cm Leichte Feldhaubitzer 18(M), but Rheinmetall-Borsig was involved in all sorts of heavy industry, as well as various aspects of small arms production.

Brandenburger Fahrrad- u. Motorrad-Werke: Brandenburger Fahrrad- u. Motorrad-Werke "Excelsior" GmbH, Brandenburg; a bicycle and motorcycle manufacturer that used the "Excelsior" tradename. Founded in 1901, it was owned by the Conrad brothers & Patz. The firm used a trademark with a stylized front bicycle wheel and an "E," which is often mistaken as "OE." During WWII, the Brandenburger Fahrrad- u. Motorrad-Werke is known to have made machinegun ammunition cans and Flak 38 magazines.

Braucke: Adolf vom Braucke AG, Gußstahl-, Draht- u. Seilwerke, Ihmerterbach b/Westig; a manufacturer of various metal goods, known to have made K98 and G43 front sight hoods during WWII.

Braun: Adolf Braun; a manufacturer of daggers and other cutlery during the National Socialist period.

Braun: Paul Braun Präzisionszieherei und Metallwarenfabrik, Esslingen a.N. (PEBRA); a supplier of high quality steel components to the small arms industry, ca. WWII.

Braun & Bloem: an ammunition manufacturer and distributor, in operation at least through the end of the 1920s. The firm stamped its cartridge casings with a "BB" within a circle.

Braunschweigische Blechwarenfabrik: Braunschweigische Blechwarenfabrik GmbH, Braunschweig; maker of RG34 cleaning kits and other sheet-metal components circa 1930-1940s.

Brehmer: Gebr. Brehmer Maschinenfabrik, Leipzig; a manufacturer of book binding machines founded in 1879 by Hugo und August Brehmer. The firm made small arms components during WWII.

Breidor: Johann Breidor, Breidora; an edged weapons maker during the National Socialist era.

Brenneke: Wilhelm Brenneke Gewehr u. Geschossfabrik, Leipzig; established in 1895 and a manufacturer of firearms and ammunition tailored especially toward the large-bore, big game hunter. After WWII the company was re-established in West Germany as Wilhelm Brenneke KG Fabrikation von Jagdgeschossen, Langenhagen.

Brno: A town in Czechoslovakia. "BRNO" has also been used colloquially to describe and market firearms made in that town or made by Ceskoslovenska Zbrojovka, Považská Bystrica, and/or Waffenwerke Brünn, Brünn under German occupation ("Brünn" being the German language version of the Czech "Brno"). See *Ceskoslovenska Zbrojovka* and *Považská Bystrica*.

Bruckmann: Ernst Bruckmann, Solingen-Ohligs; cutlery and dagger maker during the Nazi era.

BSW: Abbreviation for Berlin-Suhler Waffen- und Fahrzeugwerke, Suhl and Weimar. See *Berlin-Suhler Waffen – und Fahrzeugfabrik* and *Simson.*

Büchel: Cuno Büchel, Suhl; made sporting arms between about the 1880-1920s era.

Büchel: Ernst Friedrich Büchel, Suhl; made sporting arms marketed under the tradenames "Luna" and "Tell." Büchel was probably in business before 1900, and specialized in target rifles and free pistols. The Büchel firm survived WWII for a short time, but was eventually merged into the "VEB Meteor-Werk, Zella-Mehlis." Some reports indicate that the company may also have made components in the GDR for "VEB Merkel" and "VEB Fahrzeug- und Jagdwaffenwerk Ernst Thälmann, Suhl." See also *Thälmann.*

Bückel: Rudolf Bückel, Solingen-Merscheid; maker of cutlery and daggers during the Nazi era.

BURGO: Tradename used by H. Burgsmüller & Söhne GmbH, Kreiensen.

BURGO: Tradename later acquired by Hermann Weihrauch Gewehr und Fahrradteile Fabrik, Zella-Mehlis to market airguns and .22 caliber revolvers perhaps before, but certainly after WWII. It is also known that some small arms made after WWII by Röhm-Gesellschaft, Präzisionswerkzeug-Fabrik were also marketed under the same tradename.

Burgsmüller: H. Burgsmüller & Söhne GmbH, Kreiensen; sporting firearms maker and dealer from 19th century and beyond WWII. The company billed itself as the "largest weapons house in Germany" ca 1910. Burgsmüller used the "BURGO" and "SCHEINTOD" tradenames, among others.

Burgsmüller: Karl Burgsmüller, Berlin; of cutlery and daggers during the Nazi era.

Busch: Emil Busch AG, Rathenow; maker of compasses, photo equipment, binoculars and rifle telescopes. The origins of the Busch firm can be traced back to 1800 and the pastor, Eduard Duncker, who started making precision optics in Rathenow. Having no children of his own, Duncker sent his nephew on an apprenticeship in Berlin. Busch returned and took over the company in 1845. By 1856 the firm was producing cameras, followed by wide angle lenses in 1865. In 1872, the Optische Anstalt Rathenow went public and the name was changed to Emil Busch AG, Rathenow. Over the years, it would seem that almost an entire industry developed around Busch. In 1896 there were 163 optical businesses registered in and around Rathenow, many of which developed out of small workshops of off-duty Busch employees. During WWI, the firm made rifle scopes and binoculars for the German military. After the war, the firm survived by concentrating on the consumer market, making cameras, sporting optics, and other items. However, by 1927 the company had fallen on hard times, and it was at this point that the Zeiss firm acquired a majority stake in the legendary Rathenow company. Busch used a number of tradenames and trademarks over the decades: in the early 1900s many of its products were marketed under the "ROJA" tradename, and it also used "VISAR" to market scopes prior to WWII. During WWII,

Busch made binoculars, scopes and other optical products for the Wehrmacht. The Busch facilities were dismantled after WWII as reparations, but the remains of the company was revived by the East German government as VEB Rathenower Optische Werke and the VEB Augenoptik "Hermann Duncker." See also *Zeiss*.

Busch: Ernst Busch, Solingen; maker of cutlery and daggers during the Nazi era.

Busch: Otto Busch; an edged weapons maker during the National Socialist era. Busch used the slogan "Weltmeister."

Fig. 32: An Emil Busch AG advertisement from the late 1930s. Busch used this trademark as well as just simply the name "Busch" in script on various optics and in advertising.

Fig. 33: A Busch-made ZF41 1.5 power scope. *Courtesy Robert Jensen.*

Busch-Jaeger: Busch-Jaeger Lüdenscheider Metallwerke Aktiengesellschaft, Lüdenscheid i. Westf.; maker of components and thermoset plastics.

BVG: Abbreviation for Bochumer Verein für Gußstahlfabrikation AG. See *Bochumer*.

CE: Abbreviation for Waffenfabrik Carl Eickhorn, Solingen. See also *Eickhorn*.

Česká Zbrojovka: Česká Zbrojovka, Stakonice (renamed Böhmische Waffenfabrik AG in 1939 under German occupation); longtime manufacturer of pistols and other military products. The Czechoslovakian state was created from remnants of the Austro-Hungarian Empire after WWI. The fledgling country, however, could not supply itself with enough small arms and was forced to buy weapons from abroad or increase its own capability. To

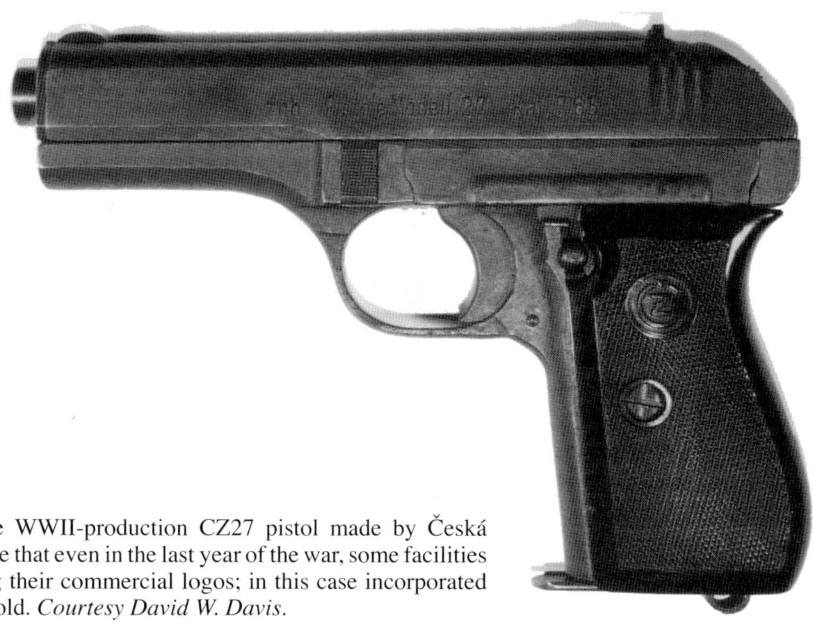

Fig. 34: A late WWII-production CZ27 pistol made by Česká Zbrojovka. Note that even in the last year of the war, some facilities were still using their commercial logos; in this case incorporated into the grip mold. *Courtesy David W. Davis.*

this end, Czech architect Karel Bubla, with backing from the Skoda Works, founded the Pilsen company "Jihoceska Zbrojovka" in 1919. The Pilsen factory at first concentrated on producing .25 caliber "Fox" vest pistols. Alois Tomiska, the designer of the "Fox," originally managed the company's production, while a new factory was being constructed in nearby Strakonice. Production was then moved in 1921 to the new facility, managed by Skoda's engineer Bartsch, with Tomiska in charge of technical management and research and development. In 1922 the firm was merged with Hubertus (J. Wenisch) into what is now known as "CZ." Expansion soon followed. In 1923, the Czech government transferred all military pistol production to CZ, and after 1926 it was the sole supplier of handguns to the Czech military. Soon thereafter CZ acquired a bicycle facility in Kralupy and then established another new facility in Uherski Brod in 1936. Work on air guns, machineguns, self-loading rifles, and flare pistols followed. When the German National Socialists under Hitler annexed Czechoslovakia in 1939, CZ cooperated with the occupiers and produced P27 pistols, P38 and MG34 components, rifles, field guns and other military products. Pistols made during German occupation can be found marked "Böhmische Waffenfabrik A. G. Prag"; Prag being the corporate headquarters of the firm. After the war, CZ was reorganized as a "people's enterprise" (Česká Zbrojovka Narodny Podnik) and continued operations. Indeed, CZ supplied thousands of refurbished German and newly-produced small arms to the Israeli state, under the direction of the post-war Zionist-leaning Czech government (until most Jews and Zionists were removed from top posts in the government via influence from Moscow). During the Cold War, CZ supplied small arms to the east, and exported commercial and police small arms abroad in the west. The company was again privatized after the end of the Cold War and is still active in the commercial and military small arms markets today.

Ceskoslovenska Zbrojovka: Ceskoslovenska Zbrojovka, Brno; Czech rifle making facility.[17] The Mauser-Werke AG, Oberndorf a/N was at the end of WWI facing a substantial loss of its assets due to the Treaty of Versailles. Instead of simply handing these over to the Allies, and ahead of any intervention from the Inter-Allied Control Commission, Mauser sold its entire well-used G98 production line to the Czechs. For their part, the Czechs gained an instant rifle production capability and Mauser generated much-needed working capital while ridding itself of machines and tooling that were due to be replaced anyway. Mauser dispatched a team led by Engineer Josef Nickl to get the line up in running in Brno, where he also established contacts in order to produce pistols of his own design under license. In 1926, Ceskoslovenska Zbrojovka acquired the Praga Zbrojovka firm, and thereafter used its trademark, a "Z" displayed inside a rifled bore. During the same period that Česká Zbrojovka, Stakonice, had succeeded in dominating the domestic pistol business, Ceskoslovenska Zbrojovka positioned itself to be the sole supplier of rifles to the Czech military. In addition, with most German companies reeling from the post-war treaty restrictions and the economic depression, Ceskoslovenska Zbrojovka became a major exporter of rifles world-wide, most notably the vz.24. The firm also developed and produced the excellent zb.26 light machinegun for the Czech military, and even allowed the British to manufacture their design under license (known as the BREN, from *Br*no, Czech city of design, and *En*field, the location of the British Royal Small Arms Factory). The facility fell under German occupation at the outset of WWII, was renamed, and operated within the

"Reichswerke Hermann Göring AG" as Waffenwerke Brünn I, Werk Brünn ("Brünn" being the German-language version of the Czech "Brno"). Its sister facility also operated under German control, as Waffenwerke Brünn II, Werk Bystrica (Považská Bystrica).[18] Both facilities made rifles for the Wehrmacht until the end of WWII. Ceskoslovenska Zbrojovka continued to make weapons after the war, but was nationalized like CZ and eventually renamed Zbrojovka Brno (of note, the use of the term "Brno" to describe Czech rifles became so wide spread that all post-war Czech firms continued to use it as a marketing tool and tradename for export sales.)[19] See also *Považská Bystrica* and *Praga Zbrojovka*.

Fig. 35: Manner in which inspectors at Ceskoslovenska Zbrojovka marked firearms made in Brno.

CeWe: Registered trademark of Carl Walther Waffenfabrik, Zella-Mehlis, which was used on molded pistol grip panels.

CGH: Abbreviation used by C.G. Haenel Waffen- und Fahrradfabrik AG, Suhl, Thüringen. See also *Haenel*.

C.G.H: Abbreviation used by C.G. Haenel Waffen- und Fahrradfabrik AG, Suhl, Thüringen. See also *Haenel*.

Chatellerault: Staatliche Waffenfabrik, Chatellerault, France, made small parts, such as followers, follower springs, rear bands and rear sight parts for the K98, G41 components, and bayonets under WWII German occupation.

Christianswerk: Christianswerk, Solingen; tooling and cutlery manufacturer, originally founded in 1824, known for dress daggers and other products during the National Socialist period. Gebr. Christians also used the abbreviated tradename "GECESO."

Fig. 36: A lengthy advertisement from the era for the Christianswerk, Solingen. This firm was famous for high quality dress daggers and other cutlery during the National Socialist era. Note the firm's distinctive "fork" trademark. *Courtesy the late William R. Smith.*

CHROMOLIT: Tradename used by Anton Wingen, Solingen to market daggers and other cutlery during the Nazi era. See also *Wingen.*

Clarfeld: Clarfeld & Co., Hemer; manufacturer of edged weapons during the National Socialist period.

Cleff: Edward Cleff, Solingen; manufacturer during the WWII period.

Clemen & Jung: Clemen und Jung, Solingen; cutlery and K98 bayonet manufacturer during the National Socialist era. The firm used a number of trademarks centered around the letter Z within a shield.

Coenders: See *Röchling, Wetzlar.*

Collath: W. Collath & Söhne, Frankfurt/Oder; a leading manufacturer of Drillings and sporting weapons from the mid-late 1800s until about 1923. The Collath firm began in 1859 when Wilhelm Collath began working for Georg Teschner's workshop. By 1870, Collath had been made a partner in the firm and took full control in the 1890s. Sons Paul and Franz took over the company when their father died in 1906 and renamed it "TESCO," a tradename used to market both firearms and proprietary ammunition. The firm apparently went out of business after the facilities were bombed by the Allies in 1941.

CONTINENTAL: Tradename used by Rheinische Metallwaren- und Maschinenfabrik, Düsseldorf to market pistols, circa before WWI. This same tradename has also been used by Spanish arms makers and many other manufacturers around the world to market various goods, such as typewriters and tires.

Coppel: Alexander Coppel, Stahlwaren- u. Waffenfabrik, Solingen, maker of cutlery and K98 bayonets. Also used the Alcoso tradename or "AWS" marking on commercial or police contracts bayonets.

Corts: Remscheider Hobelmesserfabrik Josua Corts und Sohn, Remscheid, maker of cutlery, K98 bayonets.

Crous: C.W. Crous & Co., Berga; made precision optics such as the ZF41 during WWII.

CSZ: Early abbreviation/trademark for Česká Zbrojovka (pre-WWII). See *Česká Zbrojovka.*

CULCO: Tradename used by Carl Ullrich & Co. Metallwaren, Oberschönau, bei Zella Mehlis. See *Ullrich.*

Curdtsnachf: Curdtsnachf, Solingen; marking seen on Nazi era cutlery.

CZ: Abbreviation/trademark for Česká Zbrojovka, Stakonice. See *Česká Zbrojovka.*

D: Stylized "D" logo/trademark found on pistols and other products produced by Deutsche-Werke AG, Erfurt. See also *Deutsche-Werke.*

Danzig: Königlich-Preussische Gewehrfabrik Danzig, Danzig; an Imperial German, state-sponsored small arms production facility/arsenal first established in 1817. Danzig focused on the production of M71, M71/84, Gewehr 88, Gewehr 98 rifles, and Kar.98a carbines, but also produced SG71/84 bayonets from blade blanks provided by WKC in Solingen. To keep the workforce employed after the close of WWI, the facility took the name Gewehrfabrik Danzig and made civilian hunting arms based on the familiar 98 action. In 1921 they shipped 10,000 rifle barrels to Mexico, which raised the ire of the Inter-Allied Control Commission. This resulted in the order to immediately transition to making bicycles. Then, most of the arsenal's small arms machinery, fixtures and tools were dismantled and given to the Poles to establish the Polish State Arsenal, Panstwowa Fabryka Karabinow, Warsawa.[20,21]

Fig. 37: WWI Gewehr 98 as manufactured by the Prussian arsenal in Danzig. *Courtesy Robert Jensen.*

Fig. 38: A K98 "Riemen" metal slide/adjuster made by Dransfeld & Co., Menden.

D&C: Dransfeld & Co., Spezialfabrik für Beschlagteile aller Art, Menden; manufacturer of various metal products, to include metal slides for the WWII K98 leather sling.

DECAMAR: Tradename used by Dr. F. A. Wöhler, Kassel to market optics prior to WWII. See also *Wöhler.*

Decker: Walter Decker, Zella St. Blasii; gun maker, noted for making small 6.35mm six-shot revolvers first patented in 1912.

DEKARIS: Tradename used by Carl Zeiss, Jena, to market binoculars during the 1920-30s.

DELTA: Tradename used by Hugo Linder, Solingen. See *Linder*.

DELTAR: Tradename used by Carl Zeiss, Jena, to market binoculars during the 1930s.

Deltrentis: Tradename used by Carl Zeiss, Jena, to market binoculars during the 1920-30s.

DEMAG: "DEMAG" Deutsche Maschinenfabrik AG, Duisberg; heavy equipment, dock crane and general steel manufacturer, founded in 1906. DEMAG also made ersatz bayonets in WWI and all sorts of products, such as half-tracks for the Wehrmacht during WWII. DEMAG had large subsidiaries in Berlin, Düsseldorf-Benrath, and Wetter.

DENAG: Proprietary abbreviation for Dynamit AG, Troisdorf. See *Dynamit.*

Deppmeyer: Besteckfabrik Deppmeyer, Solingen; manufacturer of edged weapons during the Nazi era.

DESIRA: Tradename used by E. Lüttges & Co., Solingen to market edged weapons during the Nazi era.

Deutsche Edelstahlwerke: Deutsche Edelstahlwerke AG, Remscheid; a maker of barrel blanks for rifles, such as the K98 and G43 in the WWII era. The firm is often abbreviated "DEW" and marked many of its products with a "E" within a hexagon.

Fig. 39: The Deutsche Edelstahlwerke trademark as stamped onto the barrel of a "duv 42" K98. "DEW" supplied barrel blanks to many rifle manufacturers during WWII. *Courtesy Leon DeSpain.*

Fig. 40: Close-up of the Deutsche Edelstahlwerke trademark: an angular "E" within a hexagon. The firm still used a variation of this same mark at the time of this writing. *Courtesy Leon DeSpain.*

Deutsche-Industrie-Werke: Deutsche-Industrie-Werke AG, Berlin-Spandau; a large corporation involved in the manufacture of motorcycles, small arms, and various other products. The company was founded on the remnants of the "Königlich-Preussische Gewehrfabrik Spandau," and several other firms and factories across Germany which were dismantled due to the conditions in the Treaty of Versailles. In December 1919 "Deutsche-Werke Aktiengesellschaft" was founded, which was in turn taken over by the "Vereinigte Industrie-Unternehmungen Aktiengesellschaft" (VIAG) in September 1922. Eventually the name Deutsche-Industrie-Werke came into use, which became the corporate parent of both the Deutsche-Werke AG, Erfurt (manufacturer of Ortgies pistols and other small arms) and Berlin (manufacturer of bicycles and motorcycles) and Spreewerk AG, Berlin (P38 pistols).[22] In addition, in the last year of WWII, D-I-W made Volksgewehr 2 receivers in Berlin, which were then assembled into complete rifles by Spreewerk's satellite plant in Grottau, Czechoslovakia, and probably another independent firm in Danzig. See also *Deutsche-Werke* and *Spreewerk*.

DEUTSCHE SELBSTLADE PISTOLE: Tradename used to market pistols made by Gotthilf von Nordheim, Zella-Mehlis in the 1920-30s.

Deutsche Waffenfabrik: Deutsche Waffenfabrik Georg Knaak, Berlin; well-known firearms manufacturer and catalog retailer in business between circa 1910-1940s.

Deutsche Waffen- und Munitionsfabriken: Deutsche Waffen- und Munitionsfabriken AG, Karlsruhe. See *DWM*.

Deutsche-Werke: Deutsche-Werke AG, Erfurt and Berlin: The initial corporate name used between 1920-1925 for the state-sponsored consortium created largely from the Imperial Amberg, Erfurt and Spandau arsenals. Deutsche-Werke fell under a larger government-sponsored consortium, the "Vereinigte Industrie-Unternehmungen Aktiengesellschaft," or VIAG. Under pressure from the Inter-Allied Military Control commission, unhappy at the pace of de-militarization within Deutsche-Werke, the consortium was ordered closed and privatized. Eventually, what remained of the Deutsche-Werke was reorganized under the "Deutsche-Industrie Werke," or DIWAG. The DIWAG eventually encompassed a number of firms and subsidiaries in several sites around Germany. After 1925, DIWAG's subsidiaries in Erfurt and Berlin retained the old name "Deutsche-Werke." During the same timeframe, Heinrich Ortgies, a German who spent many years in Liege, Belgium, designed a very practical self-loading pistol which was patented in 1918. At the end of WWI, Ortgies set up business in Erfurt and began manufacturing his design in 7.65mm for the commercial market. Ortgies manufactured some 10,000 examples carrying his "HO" trademark between about 1919-1921. Sales were brisk despite the poor state of the post-WWI economy. This drew the attention of larger companies, such as the Deutsche-Werke AG, Berlin. Sometime around 1921, the Deutsche-Werke made Ortgies a very attractive offer to buy all of his company's tradenames, inventory, components and tooling. Heinrich Ortgies accepted the proposal and apparently retired from the small arms business. Thereafter, the Deutsche-Werke in Berlin continued making bicycles and motorcycles predominantly, while its newly-acquired subsidiary in Erfurt made commercial arms and pistols marked "Ortgies Patent" but bearing the Deutsche-Werke stylized "D" trademark. See also *Deutsche-Industrie-Werke* and *Ortgies*.

Fig. 41: An Ortgies designed 6.35 mm pistol manufactured by Deutsche-Werke AG. The stylized "D" trademark was also used on bicycles and other products made by the firm. *Courtesy the late William R. Smith.*

DEW: An abbreviation of Deutsche Edelstahlwerke AG, Remscheid; a maker of barrel blanks for rifles, such as the K98 and G43, in the WWII era. See also *Deutsche Edelstahlwerke*.

Diamant: Elite-Diamantwerke, Siegmar-Schönau bei Chemnitz. See *Elite-Diamantwerke*.

Diamantwerke: Elite-Diamantwerke, Siegmar-Schönau bei Chemnitz. See *Elite-Diamantwerke*.

DIANA: Tradename used by Dianawerk Mayer & Grammelspacher, Rastatt, to market air rifles and small caliber firearms. See also "*Dianawerk*."

Dianawerk: Dianawerk Mayer & Grammelspacher, Rastatt; a maker of air rifles and small caliber firearms. The Dianawerk was founded by Jakob Mayer and Josef Grammelspacher in 1890. The firm is well-known for making air guns and training rifles and used a trademark depicting the legendary Roman huntress/goddess "Diana" holding an air rifle, with a discarded bow and arrow by her side. During WWII, the company concentrated on making stamped sheet-metal small arms components for the K98, MP44, etc. The Dianawerk and its trademark were seized by the French at the end of hostilities, and its machinery and trademark were sold to an English firm. Mayer & Grammelspacher continued on, however, maintained a close relationship with the RWS firm, eventually acquired the DIANA tradename and trademark, and was still in business at the time of this writing.

35 118596 M 12458

Fig. 42: The familiar registered trademark of the Dianawerk, showing the legendary Roman huntress/goddess "Diana" holding an air rifle, with a discarded bow and arrow by her side, as it appears in the records of the *Deutsches Markenamt.*

Dick: Friedrich Dick, Esslingen; tool and cutlery maker since ca. 1922, known for daggers and knives during the Nazi era.

Diefke: Walter Diefke, Suhl; maker of the small "KOBRA" brand pistol and a gas pistol marketed under the tradename "LACRIMAE."

Diem: Heinrich Diem Gewehrfabrik, Benshausen; a manufacturer of .22 caliber target and training rifles, circa 1920-1930s. Some of Diem's firearms were marketed under the "Zi-Di" tradename.

Dietrich: L.O. Dietrich, Altenburg; manufactured sewing machines, Sg 84/05 bayonets, machine gun components during WWI and the Louis Stange-designed FG42 during WWII.

DIOGENES: Tradename used by the cutlery maker Herder & Sohn, Solingen-Ohligs during the Nazi-era to market ceremonial dress daggers and other edgeware.

Dirlam: Ernst Dirlam Hoffnungswerk, Solingen; cutlery manufacturer during the WWII period.

Dirlam & Söhne: Dirlam & Söhne, Solingen; cutlery manufacturer during the WWII period.

DISA: Tradename/abbreviation for Dansk Industri Syndikat AG "Madsen", Kopenhagen. Originally founded as the Dansk Rekyl Riffel Syndikat A/S in 1900 to produce Madsen machineguns, this Danish firm operated under German WWII occupation. After the war DISA was known for making after-market G43 rifle magazines marked "Made in Denmark."

Dittert: J. E. Dittert J.E. & Co.; cutlery and dagger maker during the National Socialist era.

D.I.W: Abbreviation for Deutsche-Industrie-Werke AG, Berlin-Spandau. Contemporary documents also show the firm name abbreviated as "D-I-W" and "DIWAG." See *Deutsche-Industrie-Werke.*

DIWAG: Abbreviation for Deutsche-Industrie-Werke AG, Berlin-Spandau. Contemporary documents also show the firm name abbreviated as "D-I-W" and "D.I.W." See *Deutsche-Industrie-Werke.*

Dö: Abbreviation seen on barrel blanks made by Sächsische Gußstahlwerke Döhlen AG, Freital.

Dornheim: G.C. Dornheim AG, Waffenfabrik, Suhl; retailer of small arms components and small caliber training rifles. Dornheim actually had many of its firearms made under license in Spain, which were marketed with the tradename "GECADO." Dornheim was acquired in 1940 by Geco (Gustav Genschow & Co. AG, Berlin-Treptow).

Dorschel: Albert Dorschel, Solingen; cutlery maker during the Nazi era.

Dransfeld: Dransfeld & Co., Spezialfabrik für Beschlagteile aller Art, Menden (Sauerland); manufacturer of various metal products, to include metal slides for the WWII K98 leather sling.

Dr.B: See *Boehme & Co.*

Drees: Richard Drees & Sohn, Solingen; cutlery maker during the Nazi era.

Dresdner Tischfabrik: Dresdner Tischfabrik Hermann Menzel , Großröhrsdorf, bei Dresden; a furniture maker that was a principal supplier of laminated beech K98 stocks during WWII.

DREYSE: Waffenfabrik von Dreyse, Sömmerda; small arms maker. Founded by Johann-Niklaus von Dreyse in 1841, the designer and his company were famous for his revolutionary percussion needle-guns. Dreyse actually died in 1867, but his company continued on. By the late 1800s the firm had fallen on hard times, and in 1901 was bought by the Rheinische Metallwaren & Maschinenfabrik, Düsseldorf, which later became known as Rheinmetall. Since Dreyse had name recognition, Rheinische Metallwaren & Maschinenfabrik continued to use his name to market machine guns and other small arms. Indeed the top-heavy, 7.65mm, awkward-appearing "Dreyse" 1907 pistol was actually designed by Louis Schmeisser, decades after von Dreyse passed away. Rheinische Metallwaren & Maschinenfabrik produced approximately 80,000 of these pistols for the German military by 1918, and sold a few thousand more to police agencies and the Czech military after WWI. See also *Rheinmetall.*

D.R.G.M.: Not a tradename or trademark, but an abbreviation for "Deutsches Reich Gebrauchs Muster," or loosely, "German Empire Registered Design."

D.R.P.: Not a tradename or trademark, but rather, an abbreviation for "Deutsches Reich Patent," or "German National Patent."

Drilling: Term used to describe a three-barreled combination sporting arm. Not a tradename or trademark. Many firms made Drilling-type hunting weapons between 1900-1945.

Dünzer: P.A. Dünzer; an edged weapons maker during the National Socialist era.

DUO: Tradename used to market 7.65mm pistols designed by Antonin Vilímec, Kdyne, Czechoslovakia but produced by Kohout & Spol. (SPK), Kdyne, Czechoslovakia. SPK acquired Antonin Vilímec's company in 1935 and changed the marketing of his Colt 1908 clone from the "Slavia" to the "MARS"; later it was sold under the tradename "DUO."

Fig. 42a: The "DUO" tradename and logo, used to market vest-pocket pistols made by Kohout & Spol. (SPK), Kdyne, Czechoslovakia, before during and after German occupation. *Courtesy the late William R. Smith.*

Dural-Dialytan: Tradename used by Hensoldt & Söhne, Mechanisch-Optische Werke AG, Inh. Dr. Hensoldt, Wetzlar to market rifle telescopes made from a light alloy in the 1930-1940s. See also *Hensoldt.*

DÜRKOPP: Dürkopp Werke, Bielefeld; a well-respected bicycle, sewing machine, cutlery and bayonet maker. The firm was originally founded in 1867 by Nikolaus Dürkopp as "Dürkopp & Schmidt." Made bayonets and machine gun components during WWI and commercial goods between the wars. During the 1930-40s, Dürkopp made Sg 84/98 bayonets and components, and was also involved in the production of optical components during WWII.[23] The firm survived the war and made sewing machines during the Cold War, among other products.

Fig. 43: 1919 advertisement from Dürkopp, a maker of bicycles, sewing machines, and cutlery. *Courtesy Paul Shomper.*

Fig 44: The Durofol trademark as molded into a sub-contracted Karabiner 43 handguard used by Berlin-Lübecker Maschinenfabrik.

DUROFOL: Tradename used by unconfirmed company that made pistol grips, rifle handguards and other items during WWII. The Durofol trademark was last registered under Hans Joachim Gerling, Solingen-Ohligs, after WWII.

DWM: Abbreviation for Deutsche -Waffen und Munitionsfabriken AG, Karlsruhe. DWM was a huge corporation, formed by the Ludwig Loewe & Co. in order to distance itself from small arms production. In 1896 Loewe combined its arms manufacturing interests, which included its own small arms division in Berlin, Waffenfabrik Mauser, Waffen- und Munitionsfabrik A.G. in Budapest, and a 50% stake in the Belgian FN firm. With these it added the Rheinisch-Westfälische Pulverfabriken, Pulverfabrik Rottweil-Hamburg and the Deutsche Metallpatronenfabrik of Karlsruhe to form Deutsche Waffen- und Munitionsfabriken, or "DWM" as a separate organization. Loewe itself remained an independent company, still primarily involved with machine tools, and increasingly with electrical machinery during the early 20th century. DWM and all of its varied subsidiaries sold small arms and ammunition around the world and indeed, small arms sales made up a sizeable amount of Germany's total exports prior to WWI (especially to Central and South America and the Far East). During the war, DWM supplied the Imperial German Military with vast amounts of arms and ammunition, most notably P08 pistols, MG08 machineguns, and Mauser 98 rifles. After the war, DWM was no longer allowed to produce weapons by the Treaty of Versailles (although it continued to do so covertly on a much smaller scale). In 1922, the company changed its name to "Berlin-Karlsruher Industriewerke," abbreviated "BKIW" and also known as the "Berkawerke." In 1929 the Quandt Group acquired a controlling interest in BKIW, ending Loewe's involvement for good.[24] The company later added the German words "vormals" (or, "formerly") and "Deutsche -Waffen und Munitionsfabriken" to its name; but in 1936 the firm just simply went back to its old name and "DWM" abbreviation. Contracts flowed and revenue soared in the 1930s under the Nazis, and DWM quickly became one of the largest suppliers of arms and ammunition to the Wehrmacht. During WWII, DWM supervised the production of small arms and ammunition at Fabrique Nationale in occupied Belgium, controlled Mauser-Werke AG and all of its subsidiaries and had munitions plants in Berlin-Borsigswalde, Karlsruhe,

Werk Lübeck-Schlutup, and Posen.[25] Over the years, the company owned dozens of tradenames and trademarks to include DWM, Mauser, NICORRO, and others. After WWII the company structure was essentially dismantled, with portions nationalized and other subsidiaries or assets sold to other companies. In the 1950s the original Berlin subsidiary began renovating and manufacturing rail and public transport equipment, and started using the name "Deutsche Waggon- und Maschinenfabriken," but still used the original DWM trademark. The Karlsruhe branch changed its name to "Industriewerke Karlsruhe." In addition, Mauser-Werke was acquired by Rheinmetall AG and Dynamit AG, Troisdorf went on to acquire a large part of the old DWM powder and ammunition facilities, names and trademarks. See also *Loewe, Mauser and IWKA.*

DWM BB: Abbreviation for Deutsche -Waffen und Munitionsfabriken AG, Berlin. See also *DWM.*

DWM KK: Abbreviation for Deutsche -Waffen und Munitionsfabriken AG, Karlsruhe. See also *DWM.*

DWM LL: Abbreviation for Deutsche -Waffen und Munitionsfabriken AG, Lübeck. See also *DWM.*

Fig. 45: Cover page from an early 1904 DWM ammunition catalog.

Fig. 46: The DWM name and location roll-stamped on this pre-WWI Gewehr 98. *Courtesy Robert Jensen.*

Fig. 48: A nice example of a P08 Parabellum pistol, made up after WWI from leftover wartime DWM parts.

Fig. 47: A later stylized version of the DWM initials, as seen on this SMG13. *Courtesy Dr. Geoffrey Sturgess.*

Fig. 49: The same pistol as seen from above, illustrating the company's early script trademark. The numbers "1920" in this case represent the date of manufacture; but, "1920" usually denotes a weapon that was official property and controlled in accordance with the post-WWI terms dictated to Germany by the victorious Allies. "1920" is usually found in addition to the actual date of manufacture and thus should be thought of simply as a "mark of compliance."

Fig. 50: DWM's later "tonne barrel" logo, which was based on the company's heavy involvement in gunpowder and ammunition production, as illustrated in this advertisement from the 1930s.

Fig. 51: A box of DWM-produced ammunition from the same period. *Courtesy Reiner Eckert.*

DWuM: Abbreviation variant for Deutsche -Waffen und Munitionsfabriken. See *DWM.*

Dynamit: Dynamit AG, part of the I.G. Farbenindustrie group, was perhaps the largest company involved in gunpowder, propellants, and plastics production before and during WWII. The concern had a controlling interest in some forty factories and plants throughout Germany. At Troisdorf itself the company employed some 9,000-10,000 workers, of which some 2,000 were forced or relocated laborers. Of this number, about 4,500 personnel worked in the plastics division, "Venditor"; 1,000 of them engaged in machine design and engineering, essential to the operation of the facility.[26] Only about half of the company's effort was devoted to plastics manufacture, with the remainder involved in explosives and propellants research, development and production. Buildings were generally small and spread out but well staffed, equipped and tooled. Dynamit produced about 700 tons per month of Celluloid material, 1000 tons per month of phenolic molding powders, and 500 tons of urea-formaldehydes. Most of its wartime Kunststoffe production was concentrated on phenol and urea-formaldehyde based products and adhesives, but the firm is also known to have produced significant amounts of Melamine and supplied a number of other firms with tooling, machinery and molds.[27] Dynamit AG survived the war, though it was cleaved from the I.G. Farben group, and then proceeded to dominate the German petro-chemical and munitions industry over the next several decades.

E (within a hexagon): A marking used by Deutsche Edelstahlwerke AG, Remscheid. See also *Deutsche Edelstahlwerke*.

EE: Abbreviation for Erma-Waffenfabrik B. Geipel GmbH, Erfurt. See *ERMA*.

Ebel: Paul Ebel, Solingen; cutlery maker during the National Socialist era.

Ebonsmann: Ebonsmann, Solingen; manufacturer of dress daggers and other edgeware during the National Socialist period.

EBRA: A manufacturer or simply a tradename used to market 4-power rifle telescopes, circa 1920s. Little else is known about the name or the firm. Further research is clearly warranted.

ECKO: Tradename used by Emil Eckold Waffenfabrik, Suhl.

Fig. 52: Detail of a Luftwaffe-marked Modell L flare pistol and the ECKO tradename of Emil Eckold Waffenfabrik, Suhl. *Courtesy Dr. Geoffrey Sturgess.*

Eckold: Emil Eckold Waffenfabrik, Suhl; a rifle maker founded in 1876 and quite active in the 1920-1940s. Eckold is also known to have made flare pistols and sporting arms and marketed its arms under the "ECKO" trademark.

EE: Abbreviation used by Erma-Waffenfabrik B. Geipel GmbH, Erfurt. See also *ERMA*.

EICHE: Tradename and trademark used by Albrecht Kind, Berlin-Nürnberg to market small caliber target rifles made in Suhl prior to WWII. See also *Kind, Albrecht*.

Eickelberg & Mack: Eickelberg & Mack, Solingen; cutlery and dagger maker during the National Socialist era.

Eickhorn: Waffenfabrik Carl Eickhorn, Solingen; the most prolific and probably largest manufacturer of fine German cutlery, swords, daggers, knives and bayonets. First established in 1865, the company used numerous trademarks and sub-variations of the same over many decades. The firm made bayonets during WWI among other products for the military. Eickhorn was a favorite of the National Socialists. During WWII, Eickhorn predominantly made bayonets and other cutlery. Near the end of WWII the firm also made MG42 and other small arms components, as well as a small number of MP3008 SMGs based on the British Sten design. Eickhorn of course survived the war and remains in business at the time of this writing, albeit under English ownership.

Fig. 53: A 1937 Carl Eickhorn letterhead illustrating the firm's famous "squirrel" trademark.

Fig. 54: WWII-era Heer (Army) dress dagger as manufactured by Waffenfabrik Carl Eickhorn, Solingen.

Fig. 55: The Waffenfabrik Carl Eickhorn "squirrel" trademark as etched into the blade of the dagger shown on p 71.

Fig. 56: Detail of a protective bag used to ship daggers made by Waffenfabrik Carl Eickhorn, Solingen. The firm used many variations of its "squirrel" logo over the years.

Eisengiesserei-und Schlossfabrik: Eisengiesserei-und Schlossfabrik, AG Velbert; a maker of bayonets during the Imperial era. This "Iron Foundry and Lock Factory" also made padlocks.

Elite-Diamant: Elite-Diamantwerke, Siegmar-Schönau bei Chemnitz, maker of bicycles, K98 components, such as cleaning rods and rear sight components, K98 bayonets and other edgeware.

Fig. 57: 1919 bicycle advertisement for the "Diamantwerke," which at that time was a subsidiary of the larger "Elitewerke AG." *Courtesy Paul Shomper.*

Elfa: Tradename used by Elsterwerdaer Fahrradfabrik, E.W. Reichenbach GmbH, Elsterwerda-Biehla to market bicycles and commercial products. See *Elsterwerdaer*.

Elsterwerdaer: Elsterwerdaer Fahrradfabrik, E.W. Reichenbach GmbH, Elsterwerda-Biehla; a bicycle and later motorcycle manufacturer founded in 1894 by Carl Wilhelm Reichenbach. During WWI the firm made bicycles and other goods for the military, employed some 400 workers during the 1920s, and during WWII made K98 and other small arms components for the Wehrmacht. The firm marketed its commercial products under the "Elfa" tradename.

EM-GE: Tradename used by Moritz & Gerstenberger Waffenfabrik (EM-GE), Zella-Mehlis, to market flare guns, blank weapons, and air guns. After WWII, the tradename was used to market revolvers made by Gerstenberger & Eberwein, Gussenstadt, West Germany. See also *Moritz & Gerstenberger*.

Fig. 58: A newspaper illustration from the 1940s advertising EM-GE air rifles and starter pistols.

Engelbrecht & Wolff: Engelbrecht & Wolff, reportedly made pistols for the Adler Waffenwerk and marketed as the "Adler" about 1900-1910.

Englert: Englert & Solvie, Solingen; manufacturer of edged weapons during the National Socialist period.

Eppenstein: C. Eppenstein & Söhne, Solingen; cutlery and dagger maker during the National Socialist era.

Erbe: H.A. Erbe AG; an edged weapons maker during the National Socialist era.

Fig. 59: WWI Gewehr 98 as manufactured by the Prussian arsenal in Erfurt. *Courtesy Robert Jensen.*

ERFURT: Königlich-Preussische Gewehrfabrik Erfurt, Erfurt; an Imperial German state-owned arsenal that made G98s, Kar98a, P08s and other products for the military prior to and until the end of WWI. Between 1918-1920, the directors of the facility tried to keep the workforce employed by producing commercial pistols and hunting arms. The arsenal was then essentially dismantled due to the restrictions in the Treaty of Versailles. Much of what was left was absorbed by the Deutsche-Werke or transferred to Waffenfabrik Simson, Suhl. See also *Deutsche- Werke* and *Simson*.

Erfurter Maschinenfabrik: Erfurter Maschinenfabrik B. Geipel GmbH, Erfurt. See *ERMA*.

ERIKA: Tradename used by Franz Pfannl Waffen- und Patronenfabrik, Krems a/d Donau from about 1913-1926 to market small pocket pistols. See also *Pfannl*.

ERMA: Erma-Waffenfabrik B. Geipel GmbH, Erfurt; manufacturer of tooling, small arms, submachine guns, components, small caliber training rifles and sub-caliber inserts. First established in 1922 by Berthold Geipel as "Erfurter Maschinenfabrik B. Geipel GmbH," it is usually just known as "ERMA." During WWII, Erfurter Maschinenfabrik was best known for its production of the MP40, though they finished K98 barrels and manufactured many different weapons to include flare pistols, training rifles, the K98, EMP 36 (*Erma Maschinen Pistole*), tubular EMP 44, and the MP44. The region fell under Soviet occupation at the end of the war, and most of the firm's management and workers fled westward. Erfurter Maschinenfabrik was re-established in 1949 in the American Occupation Zone as "ERMA-Werke." Between 1949 and 1952 ERMA was contracted by the West German government to service and produce parts for the various weapons that had been given to the German police by the Western Allies. In 1952, ERMA established a new facility in Dachau, near München. All weapons manufactured by ERMA were produced at the Dachau facility until the company went out of business in 1998.

Fig. 60: The familiar ERMA trademark as stamped on this pre-WWII .22 caliber "Deutsches Sportmodell" training rifle. *Courtesy Robert Jensen.*

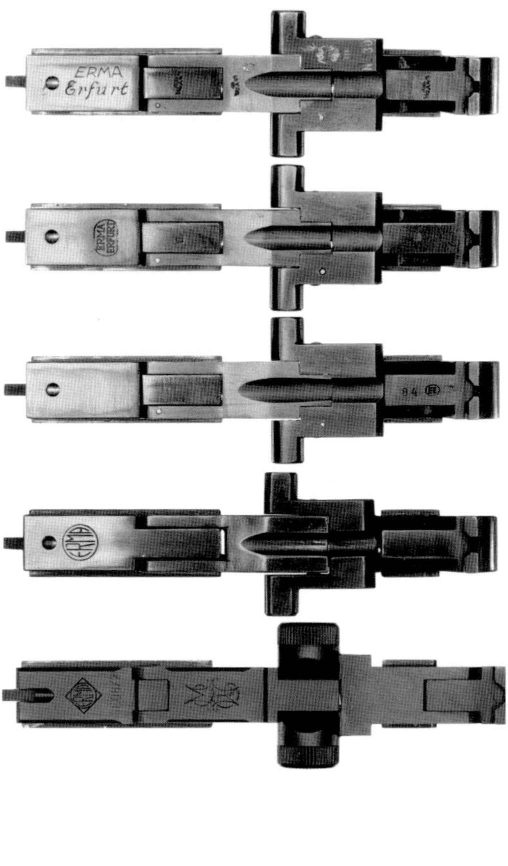

Fig. 61: ERMA used a number of trademark variations and markings over the decades as demonstrated by this group of S.E.L.f.P08 sub-caliber devices. The top two were made in the 1930s, middle ca. 1940, and the bottom two in the 1960s and 1970s (SE 08/02) respectively. *Courtesy Dr. Geoffrey Sturgess.*

Fig. 62: ERMA was reestablished in Dachau after WWII and used essentially the same trademark before and after the war, as evidenced by this post-WWII company letterhead. *Courtesy Michael Heidler.*

ERN: C. Fred. Ern Rassiermesserfabrik, Wald. Rheinl.; a maker of razors, fighting knives and other cutlery during the Imperial and Nazi eras.

Erste Nordböhmische Metallwarenfabrik: Erste Nordböhmische Metallwarenfabrik, Nieder-Einsiedel/Sud (Dolní Poustevna, Czechoslovakia); a maker of P38 pistol magazines and other sheet-metal products during WWII.

Esha: Tradename used by E. Schmidt & Habermann Jagdgewehrfabrik, Suhl, to market sporting arms circa 1920-40.

Esser: Fernando Esser & Cie, Köln; manufacturer of edged weapons during the National Socialist period.

Etablissement Charles Aufière: Etablissement Charles Aufière, Pantin (Seine), France; made precision optics and ZF41 rifle telescopes for the K98 under German WWII occupation.

Etablissement Barbier: Etablissement Barbier, Benard et Turenne, Paris; maker of precision optics and rifle telescopes under German occupation, WWII.

Everz: Arthur Everz, Solingen; maker of cutlery and daggers in the National Socialist era.

EWB: Not a trademark or tradename, but rather an abbreviation for the "Einwohnerwehr Bayern," a right-wing, conservative, anti-communist Bavarian militia which was disbanded in ca. 1921. The organization frequently identified its small arms with the letters "EWB" carved into the stocks.

Excelsior: A tradename used by Brandenburger Fahrrad- u. Motorrad-Werke "Excelsior" GmbH, Brandenburg to market bicycles and motorcycles. See *Brandenburger Fahrrad- u. Motorrad-Werke.*

Fabrique Nationale: Fabrique Nationale D'Armes de Guerre, Herstal, Belgium. Founded in 1889 by a group of Belgian investors, the company name is frequently simply abbreviated as "FN." In English, the Belgian "National Arms Factory" was established to compete with the larger Austrian and German firearms factories and state arsenals. Initially, FN concentrated on making Mauser-designed rifles under license for the Belgian Military. In 1896, the German firm Ludwig Loewe & Co., Berlin acquired majority ownership of the company (see *Loewe* and *DWM*). In the late 1890s, FN entered into a relationship with the American designer John Browning and began producing his shotgun, rifle and semi-automatic pistol designs. When Belgium was occupied by the Germans, the firm's operation was dismantled and machinery was sent to the Spandau arsenal and Waffenwerke Oberspree Kornbusch & Co. After the war, much of the machinery was returned and the firm continued operation under at first without DWM's involvement. When the Wehrmacht made short work of the Belgian military in 1940, Germany occupied Belgium again and DWM once more found itself in charge of FN. Managers and supervisors were selected from DWM or Mauser Werke AG. staff. FN employees were forced to make firearms, such as the M.1910/22, P35 and other small arms components, such as P38 components, K98 barrels, bolt bodies, cocking pieces, sears, followers and the like.[28] During this time FN was referred to in German as "DWM-Werk Lüttich, Herstal." FN produced millions of small arms over the course of several decades. After 1945 the firm concentrated on military contracts, but also remained a powerful force in the world-wide police and sporting arms market. At the time of this writing, FN still owned and marketed its products under both the "FN" and "Browning" trademarks.

Falke: Falkewerke, Wennigsen; a post-WWII manufacturer of .22 caliber training rifles, such as the "Falke Mod.36" and air rifles and pistols. Falke used a number of trademarks with a falcon as the main figurative element.

FB: Abbreviation used by Fabryka Broni W. Radomiu, Radom, Poland. See *Radom.*

FEG: Abbreviation for the Hungarian firm Femaru-Fegyveres Gépgyár R.T., Budapest. See *Femaru.*

FFG: Another abbreviation for the Hungarian firm Femaru-Fegyveres Gépgyár R.T., Budapest. See *Femaru.*

Feinbau: Feinbau Maschinen GmbH, Winterbach-Stuttgart; a metal goods and bicycle manufacturer that used the tradename "Ideal" (Jdeal, depending on the script). Feinbau is known to have made containers, canisters, tools and ammunition cans during WWII.

Feinmechanik: Feinmechanik G.m.b.H., Kassel, maker of precision optics and rifle telescopes during WWII.

Feinprüf: Tradename used by Feinprüf Feinmeß- und Prüfgeräte GmbH, Göttingen; a maker of gauges and other measuring instruments important to the National Socialist small arms industry.

Feinwerkbau: Feinwerkbau Westinger & Altenburger, Oberndorf: a post-WWII manufacturer of fine air rifles. Even though "FWB" was created in 1948, the firm is mentioned here because it was founded by two famous ex-Mauser directors, Ernst Altenburger and Karl Westinger.

Feist: A. Feist & Cie; a manufacturer of daggers and other cutlery during the National Socialist period.

Feist: A. Feist & Cie, Lunawerk; maker of cutlery and daggers in the National Socialist era.

Feist: Joseph Feist; an edged weapons maker during the National Socialist era that used the "OMEGA" tradename.

Feldbeck: Feldbeck & Pickard, Solingen; cutlery maker during the Nazi era.

Felix: Gustav Felix, Gloriawerk, Solingen; maker of cutlery and daggers in the National Socialist era.

Femaru: Tradename used by Femaru-Fegyveres Gépgyár R.T., Budapest; a Hungarian company that made pistols and the G98/40 rifle under contract for the German Wehrmacht during WWII.

Fichtel & Sachs: Fichtel & Sachs, Schweinfurt; WWI bayonet and cutlery maker. The company also made ball bearings, shock absorbers, motors, and bicycles marketed under the "Torpedo" tradename between the wars. Fichtel also produced various military goods during WWII.

Fischerwerk: Paul Kohl Fischerwerk, Solingen; a cutlery maker. The Fisherwerk tradename and trademark were first registered in 1895, but by 1925 had been acquired by Ernst Dirlam, Solingen.

Fig. 63: A 1930s advertisement from the firm Fichtel & Sachs, Schweinfurt. Note the use of the company eagle trademark in conjunction with the company's "Torpedo" brandname and trademark. In regard to this reference, Fichtel & Sachs is a known maker of bayonets and other cutlery for the German military.

F&K: Trademark/abbreviation for F. & K. Hörster, Solingen; a WWII maker of cutlery and K98 bayonets.

FLAMME: Tradename used by Louper, a WWII-era cutlery manufacturer located in Solingen.

Flocke: Flocke & Cie; manufacturer of edged weapons during the National Socialist period.

FN: Abbreviation for Fabrique Nationale d´Armes de Guerre, Herstal, Belgium. See *Fabrique Nationale.*

Ford: Ford-Werke AG, Berlin; the German subsidiary of the famous American company, Ford lost control of its German interests during the Nazi era. During WWII the firm made various automotive products, but it is also reported to have made a number of precision optics, such as the ZF41 during the era.

Förster: W. Förster, Berlin; a distributor and retailer of small arms.

Fortuna: Tradename used before WWII by J.P. Sauer/Fortuna-Werke, Suhl to market typewriters and sporting arms. After the war this company became part of the German Democratic Republic's state-owned VEB Fahrzeug- und Jagdwaffenwerk Ernst Thälmann, Suhl.[29]

Fortuna-Werke: Fortuna-Werke, Suhl, a manufacturer of sporting arms prior to WWII. After the war, this company became part of the German Democratic Republic's state-owned VEB Fahrzeug- und Jagdwaffenwerk Ernst Thälmann, Suhl.

Foss: Wilhelm Foss, Zella-Mehlis, made K98 barrels, buttplates, bands and other components which were supplied predominantly to J.P. Sauer, Suhl.

Foth: C.F. Foth, Berlin; binocular manufacturer in business ca. 1900.

Frank: Adolf Frank, Hamburg; an arms distributor-exporter in operation around 1900-1915. The firm used the abbreviated tradenames "JAGD," "GERMANIA" and "ALFA" to market various products to include tools, cutlery, air rifles, center fire rifles and shotguns.

Franke: Franke & Heideke, Braunschweig; optical firm in operation during the WWII era.

Frankonia: Frankonia Jagd Hofmann GmbH & Co. KG, Rottendorf; a post-WWII West German firearms, ammunition and accessories retailer, still in operation currently with robust catalog sales and numerous retail locations in Germany.

Frenzel: Franz Frenzel, Nixdorf; manufacturer of edged weapons during the National Socialist period.

Freund: Rudolph Freund; an edged weapons maker during the National Socialist era.

Fridericus: Fridericus, Solingen; maker's mark observed on WWII era cutlery.

Froelich: E. Froelich, Kassel; an optics manufacturer prior to WWII. Further research warranted.

Frommer: Refers to the designer of a number of Hungarian pistols, such as the "Stop" and the "Baby," while working at Femaru-Fegyveres Gépgyár R.T., Budapest, Hungary. A number of Frommer pistols were accepted by the German military in the WWI era. See also *Femeru*.

Frommia: Tradename used by Ferdinand Fromm Maschinenfabrik, Stuttgart-Bad Cannstatt. Fromm specialized in tools and precision machines.

Fuess: R. Fuess, Berlin-Steglitz; a maker of optics and especially known for making telescopes for the G98 during WWI and 10x80 binoculars for the Wehrmacht during WWII.

Funk: Christoph Funk Jagdgewehrfabrik, Suhl; a sporting arms maker established in 1835. The company specialized in Drillings throughout the 1800s. Ernst and Oscar Funk took over the concern in 1895, bought out Nimrod Gewehrfabrik, and Christoph and Alfred Funk appear to have managed the concern in the late 1930-1940s. Funk had close ties to the Römerwerk AG until the latter went bankrupt in 1931. During the latter half of WWII, the firm made MP44 and other small arms components. Funk survived the war, but what remained of the Suhl facilities were shortly thereafter merged into the "VEB Meteor-Werk, Zella-Mehlis."

Gavage: Fabrique d'Armes de Guerre Haute Precision Armand Gavage, Liège, Belgium; a small arms manufacturer operating under German occupation during WWI and WWII. A few thousand Gavage pistols were test proofed and acquired by German troops during the war.

GAZELLE: Tradename used by the cutlery manufacturer Gebrüder Krusius, Solingen during the WWII period.

GECADO: Trade name used by G.C. Dornheim AG, Waffenfabrik, Suhl and later Gustav Genschow, Berlin-Treptow to market pistols and small-caliber training rifles. Of note, the former Dianawerk also sold airguns after WWII under this tradename. See also *Dianawerk* and *Dornheim*.

GECESO: Tradename used by Christianswerk, Solingen. See *Christianswerk*.

Geco: Tradename/abbreviation used by Gustav Genschow & Co. AG, Berlin-Treptow. See also *Genschow*.

Geha: A tradename/agreed upon brandname used by several manufacturers to market converted G98 shotguns after WWI. In short, the limitations written into the Treaty of Versailles coupled with nearly non-existent military sales forced many small arms firms to find new markets and sources of revenue. One result was the oft-encountered "Geha." In 1919, the Gebrüder Rempt firm in Suhl bought up thousands of surplus G98s, converted the actions, and changed out the barrels to create affordable 12 gauge, bolt-action shotguns. These were marketed as the "Remo." In response, several other gun makers decided to collectively market a cheaper but similar version. Gustav Genschow, Gebr. Heym and Gebr. Adamy were the principles, and the tradename "GEHA" is an acronym derived from the initials of the three co-operating firms. Again, each company used surplus G98s to build up shotguns of essentially the same design, and Krupp supplied the bolt heads and barrels to all three. The original G98 stocks were sporterized and the bolt take-down disk was removed and filled with a medallion reading "Geha" in script. The Geha entered the market circa 1921-1922.

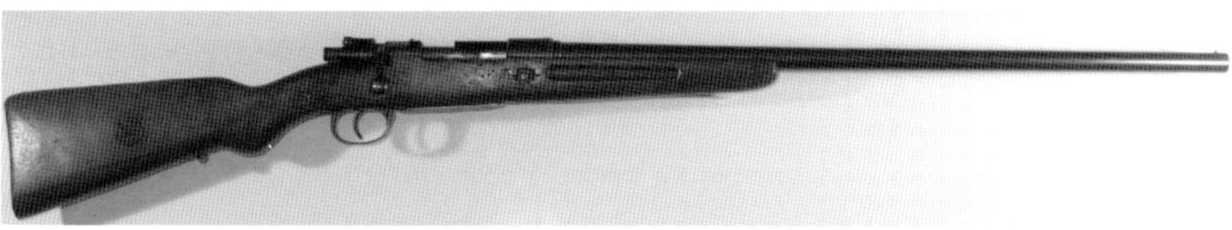

Fig. 64: The "Geha" 16 gauge shotgun, made by several firms after WWI from wartime G98 receivers.

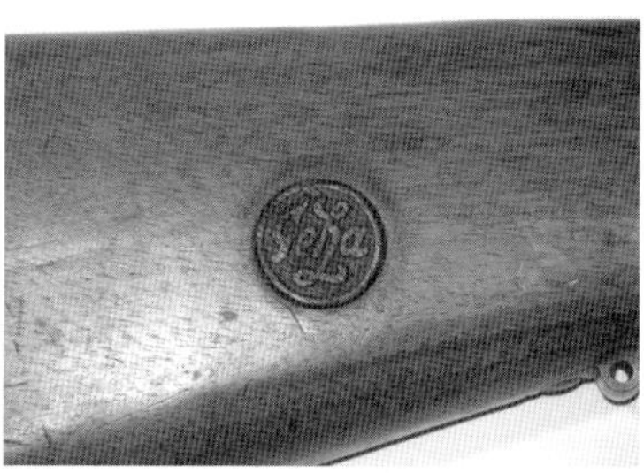

Fig. 65: The original G98 military bolt take-down stock inserts were replaced with either wooden or metal inserts bearing the "Geha" logo as illustrated above.

Geibel: Ger. Geibel, Hessische Metallwerke; an edged weapons maker during the National Socialist era.

Geigis: Friedrich Geigis, Solingen-Foche; cutlery maker during the National Socialist era.

Giesen & Forsthoff: Giesen & Forsthoff, Solingen; cutlery maker during the National Socialist era.

Geller: Robert Geller, Giessen; a maker of telescopic rifle sights from circa the 1930s which it marketed under the tradename "Uhlenflucht," probably before and after WWII.

Gembruch: Eduard Gembruch, Solingen; maker of cutlery and daggers in the National Socialist era.

GENIRA: Tradename used by Nitschke & Günther Optische Werke KG, Rathenow to market optics ca. 1920-1930s.

Genossenschaft Maschinenfabrik: Genossenschaft Maschinenfabrik, Ferlach-Kärnten, Ostmark; a WWII collective/K98 bayonet manufacturer.

Genschow: Gustav Genschow & Co. AG, Berlin-Spandau; manufacturer, distributor and importer of small arms, components, cutlery, ammunition, barrels and RG34 cleaning kits prior to, during WWII, and afterward. Genschow was founded by Dir-Ing. Gustav Genschow, and his company became a large small arms maker and catalog distributor. Genschow also was a major importer of arms for the domestic market. Genschow made ammunition at its facility in Durlach, sporting accessories, and other items in Hachenburg, and small arms at its facility in Spandau. Genschow used various oak leaf, acorn and proprietary "Geco" tradenames and trademarks during its history. It had also acquired the "GECADO" and "Kaba Special" trademarks and tradenames by at least 1929. Main products included GEHA shotguns, and cutlery, pistols, small-caliber training rifles, ammunition, and accessories. During WWII, the company made sheet-metal components for various weapons and perhaps millions of RG34 cleaning kits. Geco survived the war and was still in business at the time of this writing.

Fig. 66: A "Geco" advertisement illustrating its .22 caliber "Sportmodell" target rifle from the 1930s. Gustav Genschow & Co. AG, Berlin was a prolific manufacturer and retailer of small arms, ammunition and accessories.

Fig. 67: The receiver markings on a .22 caliber "Sportmodell" made by Gustav Genschow & Co. AG, Berlin. *Courtesy Joseph Bielik.*

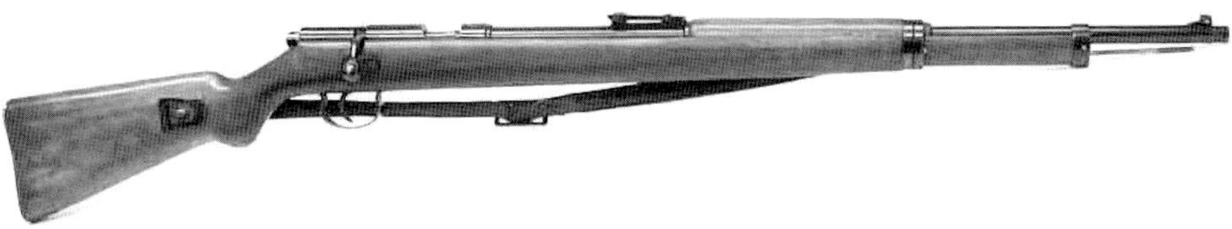

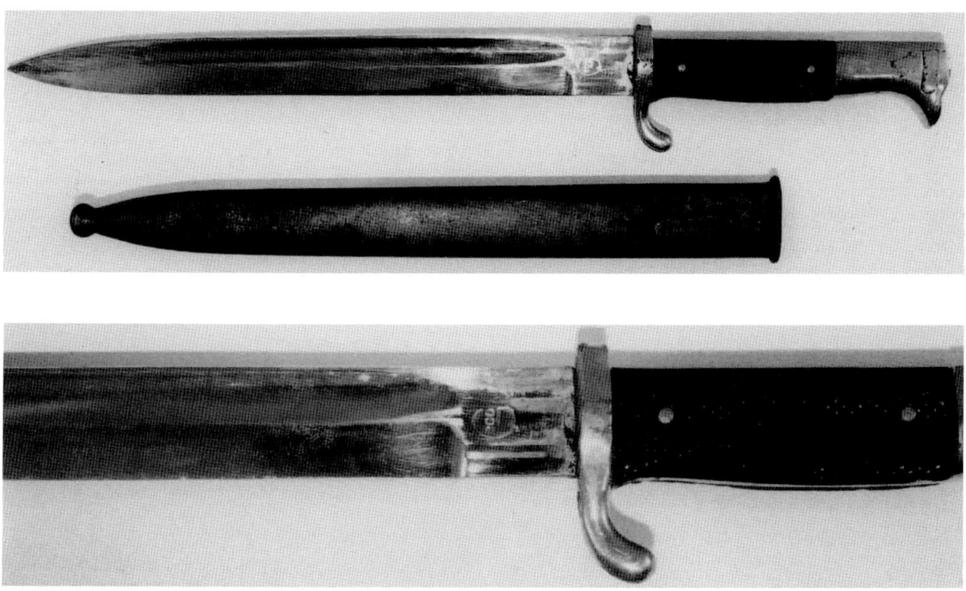

Fig. 68: A chrome-plated dress bayonet, again marketed with the "Geco" trademark.

Fig. 69: The cover of Geco's 1939 ammunition catalog.

Fig. 70: Gustav Genschow owned or controlled numerous trademarks and tradenames over several decades. The most familiar are illustrated here.[30]

Fig. 71: The markings on a 3-power WWI German sniper scope made by Gerard. *Courtesy Robert Jensen.*

Gerard: Dr. Walter Gerard, Berlin-Charlottenburg; maker of rifle telescopes circa WWI and afterward.

Gerhardt & Schubert: Gerhardt & Schubert, Amberger Kühlmaschinenfabrik, Amberg (Bayerische Ostmark); a Bavarian firm traditionally involved in the manufacture of refrigerators, but also apparently tasked to manufacture MP3008s (simplified Sten design) late in WWII.

Gering: H.M. Gering and Co., Suhl, maker of small pistols, such as the "Leonhardt" (aka "Beholla"), circa WWI.

GERMANIA: Tradename used by Adolf Frank, Hamburg, See *Frank.*

Ges. Gesch.: Not a trademark, but an abbreviation for "Gesetzlich Geschützt," or "Legally Protected" in regard to a design, tradename or trademark.

Gewehrfabrik Danzig: Name used by the former state arsenal Königlich-Preussische Gewehrfabrik Danzig to produce and market sporting arms between 1918-1920. See also *Danzig.*

Gierling: Emil Gierling, Solingen: cutlery maker during the National Socialist era.

Giersch: Robert Giersch, Solingen; cutlery maker during National Socialist era.

Giesen: Giesen & Forsthoff; an edged weapons maker during the National Socialist era.

Gladiatorwerk: Carl Wusthof (Gladiatorwerk) AG, Solingen; maker of cutlery and dress daggers during the National Socialist era.

Gloriawerk: Gustav Felix, Gloriawerk, Solingen; maker of cutlery and daggers in the National Socialist era.

Gnom: Tradename used by Optische Anstalt Oigee, Berlin, to market rifle telescopes prior to WWII. See also *Oigee*.

Gnomet: Tradename used by Optische Anstalt Oigee, Berlin, to market rifle telescopes prior to WWII. See also *Oigee*.

Göddertz: Wilhelm Göddertz; an edged weapons maker during the National Socialist era.

Goerbig: Peschakeu Goerbig, Berlin; a maker or distributor of sporting arms to include Drillings ca. 1900-1920.

Goerz: C.P. Goerz AG, Optische Anstalt, Berlin; a major manufacturer of precision microscopes, cameras, optics, scopes, and binoculars to the German military in both world wars. The facility in Berlin was acquired by Zeiss Ikon before WWII and was renamed Zeiss-Ikon AG, Optische Geräte, Goerz-Werk, Berlin-Zehlendorf.

Goerz: C.P. Goerz GmbH, Optische Anstalt, Wien; maker of precision optics and rifle telescopes. The Goerz operation in Wien was a major supplier of optics to the Austro-Hungarian armed forces in WWI. It also had a subsidiary plant in Pozsony/Bratislava that was later incorporated into the post-WWII Meopta concern. The firm used a number of tradenames over the years, such as "Certar," "Fago," "Helinox," "Pagor," "Pernox," and "Trieder." During WWII, the firm was controlled by directors appointed by Zeiss AG. See also *Zeiss* and *Meopta*.

Fig. 72: Goerz' trademark was based on a triangular prism, as seen on these Doppelfernglas 03 binoculars. *Courtesy Dr. Geoffrey Sturgess.*

Fig. 73: A 4-power sniper's telescope made by C.P. Goerz during WWI. *Courtesy Robert Jensen.*

Gold-u. Silberscheideanstalt: Gold-u. Silberscheideanstalt Oberstein, Idar-Oberstein; made stamped metal components and G/K43 rifle and other magazines during WWII.

Gothaer Waggonfabrik: Gothaer Waggonfabrik A.G., Gotha: maker of aircraft, and various parts and components.

Grabner: Georg Grabner, Rehberg bei Krems a/Donau (lower Austria); marketed and distributed Franz Pfannl's diminutive "Erika," "Kolibri," and "Kobold" miniature pistols.

Gräfrath: Gebrüder Gräfrath, Fabrik feiner Stahlwaren, Solingen-Widdert; maker of cutlery and daggers in the National Socialist era. The firm marketed some of its wares under the "Gränriso" tradename.

Grah: Carl Grah, Solingen-Ohligs; maker of cutlery and daggers in the National Socialist era.

Grah: Ernst Grah, Solingen-Foche; cutlery maker during the Nazi era.

Grah: Grah & Deppmeyer; an edged weapons maker during the National Socialist era.

Grah: Greinhold Grah; an edged weapons maker during the National Socialist era.

Grah: Gustav Reinhold Grah, Solingen-Wald; maker of cutlery and daggers in the National Socialist era.

Gran: G. Reinhold Gran, Solingen/Foch; cutlery and dagger manufacturer during the National Socialist period.

Gränriso: Tradename used by Gebrüder Gräfrath, Fabrik feiner Stahlwaren, Solingen-Widdert.

GRÄWISO: Tradename used by Gebrüder Gräfrath, Fabrik feiner Stahlwaren, Solingen-Widdert to market cutlery during the Nazi era.

Greif: Tradename used by Greifelt & Co. Gewehrfabrik, Suhl, to market sporting arms.

Greifelt: Greifelt & Co. Gewehrfabrik, Suhl; firearms manufacturer and retailer, used the tradename "Greif." Founded in 1885, the firm made K98 components, and later parts for the MP44 during WWII.

Grohmann & Sohn: Grohmann & Sohn, Metallwarenfabrik, Würbenthal/Sud; made stamped metal components as well as G/K43 rifle and other magazines during WWII.

Großfuß: Metall-und Lackierwarenfabrik Johannes Großfuß AG, Döbeln; was a key firm in the development of the MG42 and manufactured receivers and components, as well as several other innovative designs and prototypes during WWII.

Groten: Ludwig Groten, Solingen; maker of cutlery and daggers in the National Socialist era.

Grothe: Eduard Grothe & Sohn, Stendal; firearms manufacturer. Established in 1875, Grothe was granted a number of patents covering his "System Grothe," worked in cooperation with Meffert, and was in operation at least through 1925. See also *Meffert*.

Groten: Louis Groten; an edged weapons maker during the National Socialist era. They made products marked "Lanze u. Fahne."

Gründig: Karl Gründig, Dresden; a distributor and retailer of small arms.

Gunther: S. Gunther, Suhl; a maker of Drillings and high quality sporting arms in Germany prior to WWII. Gunther frequently used the marketing tradename "BOCH-DRILLING."

Gustloff: Gustloff-Werke; manufacturer of small arms, components and ammunition. The Gustloff "foundation" was formed from the Nazi takeover of several companies in the

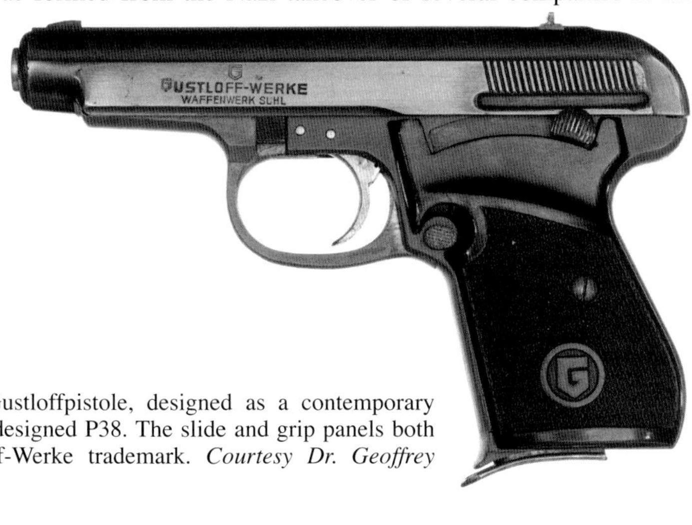

Fig. 74: A zinc-framed Gustloffpistole, designed as a contemporary competitor to the Walther-designed P38. The slide and grip panels both bear the familiar Gustloff-Werke trademark. *Courtesy Dr. Geoffrey Sturgess.*

GUSTLOFF-WERKE
FRITZ-SAUCKEL-WERK WEIMAR
RB Nr. 0/0342/0001
Anruf: Weimar Sammel-Nummer 6451, Fernschreiber 05763, Drahtanschrift: Sauckelwerk Weimar,
Thür. Staatsbank Weimar Kto. 5590, Reichsbank-Giro Weimar Kto. 76/813, Postscheck-Kto. Erfurt 18019

Fig. 75: Letterhead used by the Gustloff-Werke in the 1940s. Note the firm's Reichsbetrieb number has been pre-printed on the form. The RB was a system of national contractors identified by numbers specific to a firm, subsidiary or department.

mid-1930s, most important of which was the Jewish-owned Simson & Co., Suhl. In 1936, the Simson brothers fled Nazi Germany and emigrated to the United States. A trustee took control of the Simson assets under the framework of Dispossession of Jewish Industrialists, and then the Berlin-Suhler Waffen- und Fahrzeugwerke (BSW) was subsequently formed via mergers. In 1939, the concern was again renamed as the "Gustloff-Werke."[31] By the dawn of WWII, Gustloff-Werke had numerous subsidiaries: the Fritz Sauckel Works, Weimar (rifles, other small arms); Gustloff-Werke, Suhl (the former BSW primarily made machine guns); Otto Eberhardt Patronenfabrik, Hirtenberg, Niederdonau, Austria (formally Hirtenberger Patronen Zündhütchen & Metallwarenfabrik A.G.); another ammunition facility in Meiningen; an engineering facility in Meuselwitz; and the Branch Office, Berlin ("Thüringia House"), responsible for foreign military sales. In 1943, an additional Gustloff-supervised factory was built at the Buchenwald KZ. In August 1944, Allied bombers destroyed this "Buchenwald Werk II," killing many of the laborers. Whatever could be salvaged from the plant was reportedly moved to an underground salt mine in Billroda, where production was planned to resume. Overall during WWII, the Gustloff-Werke produced ammunition, anti-tank rifles, the MG 34/42, pistols, K98s, and G43 rifles. After the war, many of the surviving facilities were incorporated into the East German VEB Fahrzeug- und Jagdwaffenwerk Ernst Thälmann, Suhl. See also *Simson.*

Haas: Carl Haas, Solingen-Wald; maker of cutlery and dress daggers during the National Socialist era.

Haastert & Büll: R. Haastert & Büll, Solingen-Wald; maker of cutlery and dress daggers during the National Socialist era.

Habicht: Tradename used by D. Swarovski, Tyrol, Austria, to market binoculars and other optics, circa WWII and after.

HABIWA: tradename used by Hackländer & Bick. See *Hackländer & Bick.*

Hack: Josef Hack GmbH, Steyr/Ober Donau; maker of cutlery and dress daggers during the National Socialist era. Hack used a fish logo and the term "Hack-Werke, Steyr."

Hackethal: Hackethal-Draht- u. Kabelwerke AG, Hannover; a manufacturer of cable and other metal goods noted for manufacturing ammunition cans, barrel carriers and other machinegun accessories during WWII.

HACO: HACO-Werk, Ernst Mandewirth, Berlin; maker of cutlery and dress daggers during the National Socialist era.

Hackländer: Hackländer & Bick; an edged weapons maker during the National Socialist era that used the HABIWA tradename..

Haering: Eugen Haering, Solingen; maker of cutlery during the National Socialist era.

Haenel: C.G. Haenel Waffen- und Fahrradfabrik AG, Suhl, Thüringen, manufacturer of sporting and military arms, small arms components, magazines and small caliber training rifles. Founded in 1840 by Carl Gottlieb Haenel, the company was primarily involved in the production of bicycles and various machines, and then later became involved in small arms production. In 1887 the designer C.W. Aydt joined the firm and the company began making

Fig. 76: The Haenel name and location roll-stamped on this WWI Gewehr 98. *Courtesy Robert Jensen.*

Fig. 77: An MP41 cleaning brush, magazine, and loader bearing Haenel's distinctive "arrow" trademark.

target pistols. Haenel was also one of the "Suhl Konsortium" involved in making the 1888 commission rifle, and later with V. Chs. Schilling produced Reichsrevolvers. Then, Bergmann's talented designer Hugo Schmeisser eventually accepted a position within Haenel, where he continued to develop weapons, such as the MP28 and a series of vest pocket pistols. However, SMGs and other products thereafter soon dominated the firm's business and pistol production essentially ceased in 1932. Haenel had also been a small supplier of dress daggers during the Nazi era, but this too ceased in the early-mid 1930s. By the late 1930s, Haenel was still making air rifles and small-caliber target rifles, and was heavily involved in producing magazines designed by Schmeisser for MP38/40 series. The firm also manufactured K98 components, various barrels, MP28/IIs, MP40s, and MP41s. Haenel was only located about 200 meters away from the J.P Sauer facilities in Suhl, and the two firms worked closely on projects during WWII. Schmeisser's greatest achievement while working for Haenel was the development of the MP44/StG44 series, fielded near the end of WWII. After the close of hostilities, most of the Haenel machines and assets were dismantled and shipped eastward by the Russians. What remained of the company was amalgamated into the East German "VEB Ernst Thälmann Werk." See also *Schmeisser* and *Thälmann*.

Fig. 78: Haenel's familiar trademark stamped on the receiver of a pre-WWII air rifle. *Courtesy Robert Jensen.*

Hagener Gußstahlwerke: Hagener Gußstahlwerke, Remy & Co GmbH, Hagen; a maker of high quality steel, and particularly barrel blanks for the small arms industry. The company was founded in 1856, and used a number of logos, such as triangle with tools, an "HR" under a crown, and marketed many of its products under the "Presto" tradename.

Fig. 79: A 1905 advertisement from the Hagener Gußstahlwerke, Remy & Co GmbH, Hagen. Note the "crown over HR" trademark.

Hahn: A.-G. Hahn, Cassel [Kassel]; made ballistic testing equipment and rifle telescopes for the military and commercial markets circa WWI.

Hahn: Hermann Hahn, Solingen-Wald; maker of cutlery during the National Socialist era.

Hahn & Kolb: Hahn & Kolb, Werkzeugmaschinen-Fabrik, Stuttgart; a toll maker noted to have made various small arms broken case extractors, armorer's tools and inspection gauges during WWII. Hahn marked its tools "HK," "H&K," "H.u.K." and "eoz." See also *Orion*.

Häker: Gustav Häker; Solingen; maker of cutlery during the National Socialist era.

Halbach: Carl Halbach, Solingen; maker of cutlery, daggers, knives et al during the National Socialist period.

Halbach: Gebr. Halbach, Weyer b.Ohligs; made cutlery and dress daggers marketed with its "bulldog" trademark, circa 1930s.

Halbach: Wilhelm Halbach, Solingen; maker of cutlery, daggers, knives et al during the National Socialist period.

Halbe & Gerlich: Halbe & Gerlich, Hamburg; manufacturer of arms and ammunition in the 1920s. The firm used the tradename "HALGER," an abbreviation/compression of the first three letters of the two company names.

HALGER: Tradename used to market arms and ammunition by Halbe & Gerlich, Hamburg.

Halverscheid: H. Halverscheid, Hagen; made shotguns, Drillings and other sporting arms ca. 1900.

Hammesfahr: Hammesfahr & Cie, Solingen; cutlery and dagger maker during the Nazi era.

Hammesfahr: Gottlieb Hammesfahr Stahlwarenfabrik AG, Solingen-Foch; made a variety of military goods including knives, bayonets, and daggers during the WWI-WWII period.

Hartkoop: Rich. & E. Hartkoop; an edged weapons maker during the National Socialist era.

Hartkopf: Hartkopf & Co, Solingen; cutlery and dagger maker during the WWI through Nazi era. Hartkopf used a number of variations of its "Double Devil" trademark over the years.

Hartmann: Hartmann-Optik Wetzlar, Werkstätten für Feinoptik und Feinmechanik Inh. Karl Hartmann, Wetzlar; confirmed to be a post-WWII manufacturer of binoculars, but presently not known to have made optics prior to the end of WWII.

Hartmann: Robert Hartmann; an edged weapons maker during the National Socialist era.

HASAG: Tradename of Hugo Schneider AG. HASAG was a major corporation involved predominantly in the manufacture of gunpowder, propellants and explosives. During WWII, HASAG also made LP42 flare pistols and the Panzerfaust. HASAG had subsidiaries in Berlin-Köpenick, Altenberg, Dermbach, Eisenach, Kielce (Poland), Langenwiesen, Meuselwitz, Taucha, Oberweißbach, and Leipzig.

Hauptner: H. Hauptner, Berlin; an edged weapons maker during the National Socialist era.

Hauswirtschaftsmaschinen-Gesellschaft: Hauswirtschaftsmaschinen-Gesellschaft mbH, Berlin-Tiergarten, Rg34 cleaning kit manufacturer, used the tradename "HAWIG."

HAWIG: Tradename/abbreviation of Hauswirtschaftsmaschinen-Gesellschaft mbH, Berlin-Tiergarten; manufacturer of RG34 cleaning kits and other military products.

Fig. 80: Progression of HAWIG (Hauswirtschaftsmaschinen-Gesellschaft mbH, Berlin-Tiergarten) made RG34 cleaning kits. In the 1930s, the firm used their traditional trademark, then began adding dates to the marking, and finally began using the "cmr" military letter code during WWII.

Hebel: Not a tradename, but rather a 26.5mm flare pistol design manufactured by several different companies from 1894 onward. "Hebel" translates as "lever," referring to an "under lever" action. WWI-era Hebel flare pistols have been observed with markings from several manufacturers, to include: J.G. Anschütz, Germaniawaffenfabrik, Zella-Mehlis, "CHR.F," "B&E," "Gebr. R." and "PL."

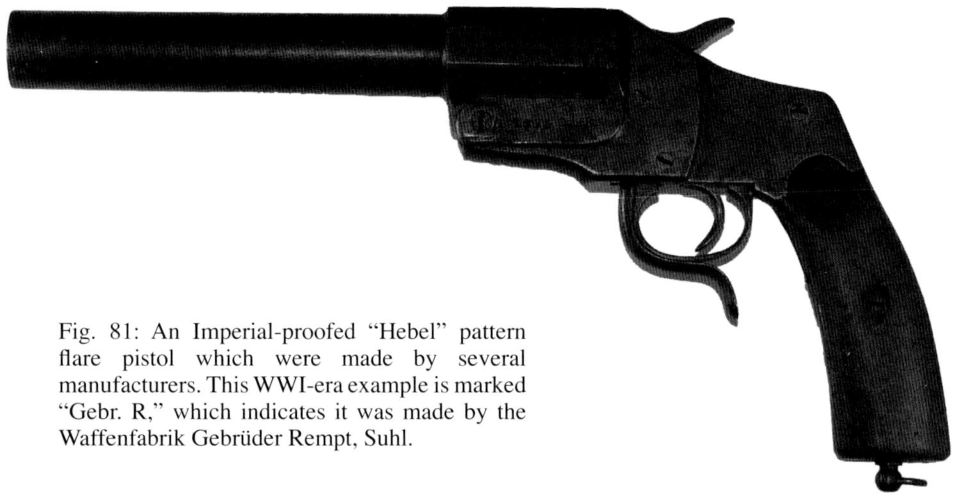

Fig. 81: An Imperial-proofed "Hebel" pattern flare pistol which were made by several manufacturers. This WWI-era example is marked "Gebr. R," which indicates it was made by the Waffenfabrik Gebrüder Rempt, Suhl.

Heele: Hans Heele, Berlin; manufactured rifle telescopes for the G98 circa WWI.

Heim: Tradename used by C.E. Heinzelmann, Plochingen a/Neckar. See *Heinzelman*.

Heinemeyer: Heinemeyer; designed and manufactured an inexpensive 5mm emergency pocket pistol circa 1930-1940s.

Heinzelmann: C.E. Heinzelmann, Plochingen a/Neckar; maker of the 6.35mm "Heim" pistol between about 1910-1920s. Mauser-Werke later acquired the rights to this design and manufactured a slightly modified version which became known as the "WTP I" and WTP II."

Heindelberg: Carl Heindelberg; Solingen; cutlery and dagger maker during the Nazi era.

Helbig: F. & A. Helbig (Frieda & Alfred Helbig), Steinbach bei Schleusingen; maker of cutlery, daggers, knives et al during the National Socialist period.

Helfricht: Tradename used to market pistols made by A. Krausser, Zella-Mehlis. See also *Krausser*.

Helfricht & Fischer: Helfricht & Fischer, Zella-Mehlis; manufacturer of sporting arms circa 1933-40.

HELIA: Tradename used on rifle telescopes made by Karl Kahles, Optiker, Wein.

HELIAACHT: Tradename used on 8-power rifle telescopes made by Karl Kahles, Optiker, Wein.

HELIASECHS: Tradename used on 6-power rifle telescopes made by Karl Kahles, Optiker, Wein.

HELIAVIER: Tradename used on 4-power rifle telescopes made by Karl Kahles, Optiker, Wein.

HELKRA: tradename used by A. Krausser, Zella-Mehlis.

Hella: Tradename used by Runge & Kaulfuss, Rathenow to market binoculars and opera glasses. A manufacturer of bulbs and other electrical equipment was also named Hella during the WWII period, but is not relevant to this text. See also *Runge & Kaulfuss.*

Heller: Gebr.Heller, Marienthal bei Schweina; maker of cutlery, daggers, and bayonets WWI-WWII period.

Heller: Gebrüder Heller, Schmalkalden; tool and cutlery maker during the Nazi era.

Hemming: Edmund Hemming Fabrik für Fahrrad-, Motorrad- u. Automobilteile, Benshausen; bicycle and auto parts maker founded in 1905, made small arms components among other items during WWII.

HE-MO: a brand name of 7.65mm pistol made and marketed in Germany, manufacturer unconfirmed at the time of this writing.

Henckels: Henckels AG, Zwillingswerk, Schneidwarenfabrik u. Stahlwerk, Solingen. Henckels' "Zwillinger" or "twins" trademark is one of the oldest in existence, first registered by Paul Henckels in 1731. By 1818 the firm had opened a branch office in

Fig. 82: An early advertisement for the Henckels AG, Zwillingswerk, Solingen-showing the company's distinctive "Twins" trademark. *Courtesy the late William R. Smith.*

Fig. 83: Henckels even had the firm's trademark incorporated into the brick work of the factory as seen in another advertisement. *Courtesy the late William R. Smith.*

Berlin, followed by sales offices in New York, Wien, Paris, Hamburg, Dresden, München and Köln. Henckels was the first manufacturer in the area to use steam-powered forges and mechanized hammers for drop forging. Over the many years, Henckels won numerous guild and industry awards for its products. During the National Socialist era, Henckels further expanded its reputation as a maker of fine scissors, knives and dress daggers. After WWII, the firm acquired the assets and trademarks of several smaller firms and was still in business at the time of this writing.

Henkel: Carl Henkel, Bielefeld; cutlery manufacturer, known for dress bayonets during the 1930s.

Henkel & Müller: Henkel & Müller, Solingen; a cutlery maker that used the "Macero" tradename to market dress daggers during the National Socialist period.

Henkels: Paul Henkels, Solingen; cutlery maker during the Nazi era.

Henschel: Henschel & Sohn GmbH, Kassel; a maker of locomotives, airplanes and the Tiger tank during WWII. The company was founded in 1810 by Georg Christian Carl Henschel. In regard to the subject of this reference, Henschel was also involved in Volksgewehr production near the end of WWII.

Hensoldt: Hensoldt & Söhne, Mechanisch -Optische Werke AG, Inh. Dr. Hensoldt, facilities located in Wetzlar and Herborn. The company was founded by brothers-in-law Moritz Carl Hensoldt and Carl Kellnerin in 1849. However, by 1850 Hensoldt had broken away and started his own company "M. Hensoldt & Söhne AG," concentrating on the manufacture of optical instruments. The firm was in business for many decades, though Hensoldt himself died in 1903. The company prospered by supplying rifle telescopes and binoculars to the Kaiser's military during WWI. The company survived the war and the post-war depression, but ultimately was taken

Fig. 84: An early post-WWI advertisement from the Hensoldt firm illustrating the company trademark and the "Dialyt" tradename. *Courtesy Paul Shomper.*

Fig. 85: A 1920-1930s shipping box for a 2 3/4 power scope used with small bore rifles, which bears both of Hensoldt's familiar trademarks.

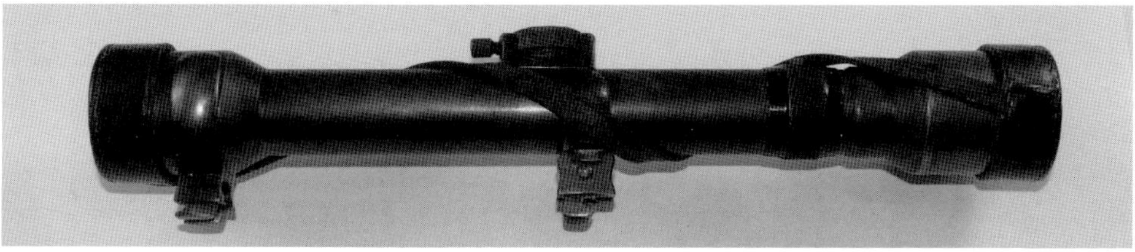

Fig. 86: 1930s-era Hensoldt 4-power scope which has been engraved with the company's best-known prism-type trademark.

over by Carl Zeiss AG, Jena in 1928.[32] Subsequent to this, the two firms shared designs, technology and improvements, for instance switching from brass to lighter metals, such as zinc and aluminum. Hensoldt continued making sporting and military rifle telescopes, binoculars and other optics through the end of WWII and beyond. The firm used a number of tradenames over the years to include "DIAPI," "DIASTA," "DIALYT," "DIALYTAN," "DURAL-DIALYTAN" (post-WWII) and others. After the war, the Zeiss tended to use the Hensoldt name for domestic military sales and the Zeiss name for export and commercial sales. See also *Zeiss*.

Herbeck & Meyer: Herbeck & Meyer, Solingen-Höhscheid; a cutlery maker during the National Socialist period.

Herbertz & Meurer: Herbertz & Meurer, Solingen-Gräfrath; cutlery maker during the Nazi era.

Herder & Engels: Herder & Engels, Solingen-Ohligs; cutlery and dagger maker during the Nazi era.

Herder: Friedrich A. Herder und Sohn, Solingen; maker of cutlery, daggers, K98 bayonets 1930-1940s.

Herder: H. Herder, Solingen; cutlery maker during the National Socialist period.

Herder: Richard A. Herder, Solingen; maker of cutlery, daggers, and bayonets during the Nazi-era and beyond. The firm made Sg98/05 bayonets during WWI and Sg84/98 bayonets during the Nazi era. The firm also made MG34 top covers, grenade launchers, G43 triggers, and MP44/StG44 components during WWII.

Herder: Robert Herder, Solingen; maker of cutlery and daggers during the Nazi-era and beyond.

Herder: Herder & Sohn, Solingen; maker of cutlery and daggers during the Nazi-era, frequently marketed with the "Diogenes" tradename.

Herkenrath: Friedrich Herkenrath, Merscheid; manufacturer of edged weapons during the National Socialist period. The company used the "Ben Hur" tradename.

Hermes: Emil Hermes, Solingen; manufacturer of edged weapons during the National Socialist period.

HEROLD: Tradename used on firearms and rifle telescopes made and/or distributed by Franz Jäger & Company, Suhl.

Hertel & Reuss: Hertel & Reuss, Optik u. Feinmechanik, Kassel; maker of rifle telescopes, binoculars and other optics from 1927-1995. After WWII, the company was acquired by B. Nickel, Marburg.

Herz: Herz, Berlin; maker of binoculars and other precision optics (era unconfirmed).

Hessische Industrie-Werke: Hessische Industrie-Werke, Wetzlar. See *Röchling*.

Heym: Friedrich W. Heym, Suhl; a sporting arms maker. The company was founded in 1865 by the company namesake. In 1912, Adolf Heym assumed management of the firm. From ca. 1933-1945 the company made small caliber sporting and training rifles. After the post-WWII Soviet takeover of the Suhl area, the firm was reestablished in West Germany as Friedrich Wilhelm Heym GmbH & Co., Münnerstadt.

Heym: Geb. Heym, Suhl; manufactured fine sporting arms during the 1920-1930s.

H.H.K: See *Klüssendorf*.

Hirtenberg: Hirtenberger Patronen- und Zündhütchenfabrik, Hirtenberg, Niederdonau, Österreich (Austria); later renamed the Otto Eberhardt Patronenfabrik, it was a manufacturer of ammunition and cartridges before, during, and after WWII. Hirtenberg had munitions factories in Hirtenberg, and another in Ronsdorf bei Wuppertal. The company was absorbed into and supervised by the Gustloff-Werke, Suhl prior to the opening of WWII. After the war, the firm was reestablished and continued to produce ammunition for the military, police agencies and sportsmen.

Hohner: Matthias Hohner AG, Harmonikafabrik, Trossingen; a harmonica manufacturer since the 1800s that was pressed into making fuses and other munitions components during WWII. A number of sound suppressors/silencers marked with the firm's "edb 44" ordnance code have surfaced and are thought to have been intended for the G/K43 rifle. They may have been made by a clandestine government agency, though, after WWII and marked in this manner to cloak the original point of origin. At least thirty of these aluminum body, black-painted examples are known to exist. Further research is clearly warranted.

Höller: F.W. Höller & Co. Waffenfabrik, Solingen; maker of cutlery and bayonets during WWI and the Nazi era.

Hommel: Hommelwerke GmbH, Mannheim-Käfertal; a tool maker noted for the production of various small arms tools and manufacturing gauges during the National Socialist era.

Homrich: Ferdinand Homrich & Sohn, Hamburg: an optical firm in operation during the WWII era.

Hoppe: Curt Hoppe, Solingen; cutlery maker during the National Socialist period.

Hoppe: Gottfried Hoppe & Söhne, Solingen; cutlery maker during the National Socialist period.

Hoppe: Robert Hoppe; an edged weapons maker during the National Socialist era.

Hoppe: Wilhelm Hoppe, Inh. Helmut Giesen, Solingen; cutlery and dagger manufacturer during the National Socialist era.

Hörster: E&F Hörster Stahl- u. Metallwarenfabrik, Solingen; maker of cutlery and bayonets prior to WWI and through the National Socialist era and WWII.

Hörster: F.& K. Hörster, Solingen; maker of cutlery and K98 bayonets during WWII.

HUBERTUS: tradename used by Cuno Ritter as well as J. Albert Schmidt to market cutlery from ca. 1900. See also *Ritter* and *Schmidt*.

HUBERTUS: Tradename used by Albe Imman. Meffert Gewehrfabrik, Suhl to firearms from about 1839-1945. Products with this tradename were also marketed by AKAH (Albrecht Kind, Berlin-Nürnberg) to include scopes, scope cases, and Hitler Youth knives. Additionally, scopes were also marketed under the Hubertus brandname, thought to have been manufactured by Karl Kahles, Optiker, Wein, perhaps prior to 1945, and then for certain after WWII by Schmidt & Bender, Biebertal/Wetzlar.

HUBEO: Tradename used by Hugo Berns, Solingen-Ohligs to market daggers and other cutlery during the National Socialist regime.

H.u.K.: Abbreviation for Hahn & Kolb, Werkzeugmaschinen-Fabrik, Stuttgart. See *Hahn & Kolb*.

HW: Abbreviation and tradename of Hermann Weihrauch Sportwaffenfabrik on revolvers made in West Germany after WWII.

HWA: HW Abbreviation and tradename used by Herman Weihrauch Sportwaffenfabrik after WWII. See also *Weihrauch*.

H. W.-Z: Abbreviation for Hermann Weihrauch Gewehr und Fahrradteile Fabrik, a small arms and accessories manufacturer in Zella-Mehlis prior to the end of WWII, and West Germany after the war. See also *Weihrauch*.

Ideal: Tradename used by Feinbau Maschinen GmbH, Winterbach-Stuttgart. See *Feinbau*.

I.G. Farben: Industrie Gesellschaft Farben was originally formed in 1925 by merging the Bayer, AGFA, BASF, and Hoechst corporations with two smaller companies.[33] It eventually grew into a multi-national conglomerate of industrial cartels, with a controlling interest in over 380 German and over 500 foreign firms. In 1933, the conglomerate signed a contract with the National Socialist regime to supply virtually all of the petrol and fuel needed by the German armed forces. In addition, I.G. Farben produced all of Germany's synthetic rubber and lubricating oils, and a large portion of the country's steel, light metals, explosives, plastics, and chemicals. I.G. Farben is also infamous for its production of Zyklon-B (originally developed as a pesticide, but used to execute prisoners in the extermination camps) and the nerve agent Tabun. Dr. Otto Ambros headed the company during WWII, which by 1943 was probably the largest corporation in the world. In the context of this reference,

Fig. 87: An I.G. Farben advertisement from a 1937 periodical. The concern also used a stylized lower-case "ig" logo, and by 1943 was probably the largest corporation in the world.

I.G. Farben was the corporate parent of AGFA Kamerawerk, a maker of ZF4 telescopes for the Wehrmacht. After the war I.G. Farben was seized and dissolved by the Allies' Control Commission Law Nr. 9, and many of the high-level managers served prison time. The original primary subsidiary companies remain in business of course today.

Inhauser: Carl Inhauser, Wien; manufacturer of military small arms in Vienna, Austria, from about 1920-1945.

INOX: not a brandname, but rather an international designation for "stainless steel"; frequently seen on WWI-WWII era German cutlery.

IWKA: Abbreviation for Industriewerke Karlsruhe AG, Karlsruhe; the post-WWII incarnation of Deutsche -Waffen und Munitionsfabriken AG, Karlsruhe. See also *Loewe, Mauser and DWM.*

Jackenkroll: Adolph Jackenkroll Optische Anstalt, Berlin; manufacturer of rifle telescopes. The firm was founded in 1879, supplied telescopes to the German Army in WWI, and then in WWII made and marketed optics with the "AJACK" tradename. A large number of AJACK scopes were supplied to neutral Sweden in exchange for high-quality Swedish steel. Substantial numbers were also supplied to the Waffen-SS.

Jacobs & Co.: Jacobs & Co., Solingen-Gräfrath; cutlery maker during the National Socialist era.

Jacobs: Rudolf C. Jacobs, Solingen-Gräfrath; cutlery maker during the National Socialist era.

JAGA: Tradename used by Frantisek Dusek on pistols made in Opocno, Czechoslovakia, in about 1930.

JAGD: Tradename used by Adolf Frank, Hamburg. See *Frank.*

Jäger: Jäger & Co, Silberwarenfabrik, Kaiserwerth; an edged weapons maker during the National Socialist era.

Jäger: Franz Jäger & Co. GmbH, Suhl; gun maker and firearms retailer from around at least 1910 through to the end of WWII. The company was the maker of the stamped sheet-metal 7.65mm "Jäger Pistole," and also used the "Herold" tradename on many of its products.

Jaeger: Richard Jaeger, Gewehrlauffabriken, Suhl; barrel maker in the 1930-1940s.

Jaguar: Tradename used to market rifle telescopes made by W & H Seibert, Wetzlar.

Jenisch: Oskar Jenisch; an edged weapons maker during the National Socialist era.

Jenoptik: Jenoptik, Jena; a post-WWII optical firm, part of the VEB Zeiss, Jena, group, GDR. See also *Zeiss*.

J.G.A.: Abbreviation/trademark of J.G. Anschütz, Germania Waffenfabrik, Zella-Mehlis. See *Anschütz*.

Johanniswerk: Johanniswerk, Bayreuth (a.k.a. Johnswerk); an edged weapons maker during the National Socialist era.

Jopp: Andreas Jopp, Zella-Mehlis; manufacturer of bicycles, small arms components and other metal ware, in business between 1908-1948. Jopp was eventually combined with other small firms in and around Zella-Mehlis into the "VEB Meteor-Werk, Zella-Mehlis."

Jordan: F. W. Jordan, Solingen; cutlery maker during the National Socialist era.

JOS.CORTS: Abbreviation used by Remscheider Hobelmesserfabrik Josua Corts und Sohn, Remscheid; a maker of cutlery and bayonets during WWI and WWII.

Jostes: Jostes & Co.; an edged weapons maker during the National Socialist era.

J.Sch.: Abbreviation used by the cutlery and bayonet manufacturer Julius Schmidt und Söhne, Riemberg bei Breslau, Silesia.

JULANCO: Tradename used by J. Langenberg & Co to market edged weapons during the National Socialist era.

Juliuswerk: Juliuswerk- J. Schmidt & Sohn, Riemberg bei Breslau, Silesia; a maker of cutlery during the Nazi era.

Jung: August Jung, Viernau bei Suhl; manufacturer of Drillings and sporting weapons.

Just: Josef Just, Ferlach; a manufacturer of sporting arms in Ferlach, Austria, 1920-1930s.

Justinus: Justinius Werk, Solingen; cutlery manufacturer during the National Socialist period.

K55K: Trademark used by Heinrich Kaufmann & Söhne, Solingen. See also *Kaufmann*.

Kaba: Tradename originally used by Karl Bauer & Co., Suhl, a retailer and distributor of firearms, many of which were made elsewhere (such as August Menz) and simply carried the "Kaba Spezial" brandname. Gustav Genschow & Co. AG, Berlin-Spandau had acquired the trademark and name by ca. 1929.

Fig. 88: a "Liliput" made by August Menz of Suhl, but marketed under the "Kaba Spezial" brand. *Courtesy the late William R. Smith.*

Kahles: Karl Kahles, Optiker, Wein; a maker of precision optics and rifle telescopes. Karl Robert Kahles formed the company in 1898 by buying and merging the smaller Simon Plössl firm and the Optischmechanische Werkstätte Karl Fritsch. After the founder died in 1908, the company was taken over by Ernst and Karl Kahles. By 1900, Kahles was producing precision telescopes, binoculars and other optical equipment for the military. The firm introduced the Helia series of telescopes in 1926. Kahles marketed rifle telescopes under the Helia, Mignon, Argus, Telorar, and Hubertus tradenames. The firm survived WWII and was still producing quality optics at the time of this writing.

Kaiser: Emil Kaiser und Co., Solingen; cutlery maker during the National Socialist era.

Kaiser: Max Kaiser Waffenhammer, Deggendorf; cutlery maker during the National Socialist era.

Kaldenbach: Karl Robert Kaldenbach, Solingen; cutlery maker during the National Socialist era.

Kallenbach: Kallenbach, Meyer & Franke, Metallwarenfabrik, Luckenwalde; a metal goods manufacturer headed by Ernst Meyer, noted for making ammunition cans and Flak38 magazines during the WWII era.

Kamphausen: Kamphausen & Plümacher, Solingen-Ohligs; an edged weapons maker during the National Socialist era.

Kaps: Karl Kapps, Wetzlar; a manufacturer of rifle telescopes after World War II.

23 716606* St 3948

KASTO

Fig. 89: The registered trademark of the Kasto Maschinenbau GmbH & Co. KG., Achen as it appears in the records of the Deutsches Markenamt. Kasto had little to do with firearms production with the exception of VK98 stocks near the end of WWII.

KASTO: Kasto Maschinenbau GmbH & Co. KG., Achen; maker of VK98 rifle stocks near the end of WWII. KASTO is an abbreviation of the founder's first and last name, Karl Stolzer. The firm began as a carpentry shop in 1844 and gradually grew into a successful manufacturer of tools and saw blades. KASTO was still in business at the time of this writing.

Kastor: Kastor & Co., Solingen-Ohligs; a manufacturer of edged weapons during the Nazi era.

Kaufmann: Heinrich Kaufmann & Söhne, Indiawerk, Solingen; a cutlery firm especially known for their daggers and handy pocket knives, which were popular with German soldiers before, during, and after WWII. Kaufmann used the "Mercator" tradename and the familiar cat over "K55Ꮖ" trademark. The firm also made Sg 98/05 bayonets in cooperation with L.O. Dietrich during WWI.[34]

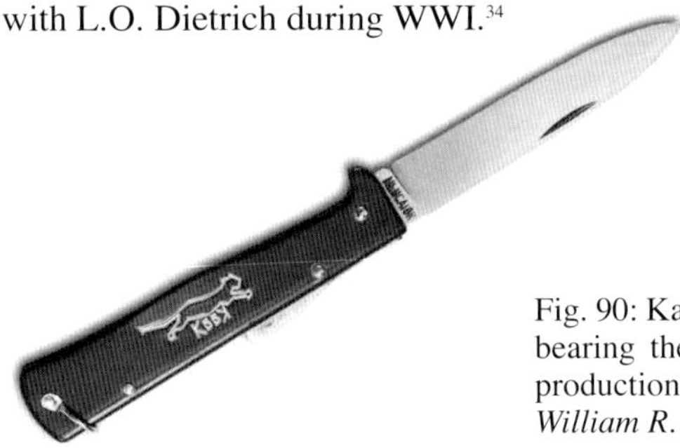

Fig. 90: Kaufmann's familiar "Mercator" brand pocket-knife bearing the "Cat K55K" logo. This style of knife was in production for many decades (and still is). *Courtesy the late William R. Smith.*

Kayser: C.F. Kayser, Solingen; maker of cutlery and dress daggers in the 1930-40s.

Kemper: Ernst Kemper, Solingen; cutlery maker during the National Socialist era.

Kern: Kern, Kläger & Cie. Lederwaren, Berlin; a leather goods maker, particularly noted for holsters and binocular cases during WWII.

Kerner: Kerner & Co., Gewehrfabrik, Suhl; small arms maker founded in 1936. Kerner originally made Drillings and then produced small arms components during WWII. The company appears to have ceased operations at the close of the war. Also known as Gewehrfabriken Emil Kerner & Sohn.

Kessler: F.W. Kessler, Suhl; maker or retailer of early auto-loading pistol. This firm was associated in some way with Friedrich Pickert, Zella-Mehlis.

Kettner: Eduard Kettner, Suhl and Köln; custom Drilling and sporting arms maker/ distributor circa 1920-1930s to the modern day.

KH: Abbreviation for K. Helfricht, a designer of small vest pocket pistols manufactured by A. Krausser, Zella-Mehlis.

Kind: Albrecht Kind, Berlin-Nürnberg; a large catalog retailer and maker of holsters, scope cases, hunting wares, and leather goods under a number of different tradenames and trademarks (also see *Akah*, *AKAH* and *HUBERTUS*). The firm distributed and sold most brands of commercially available ammunition, blank, air and cartridge firearms of the era and operated much like the modern-day American firm Cabellas, Inc. or the German Frankonia Jagd GmbH.

Fig. 91: One of Albrecht Kind's early trademarks as it appeared in the records of the *Deutsches Markenamt*.

Fig. 92: The cover of Albrecht Kind's 1932 "Akah" retail catalog, demonstrating both the firm's tradename and one of its many trademark variations.

Kirschbaum: F.A. & Co Kirschbaum & Co., Solingen; cutlery maker during the National Socialist era.

Kirschbaumer: Georg Kirschbaumer; an edged weapons maker during the National Socialist era.

Kirschner: Julius Kirschner & Sohn; an edged weapons maker during the National Socialist era.

Klittermann & Moog: Klittermann & Moog GmbH, Haan b/Solingen; cutlery maker during the Nazi era.

Klaas: Robert Klass Messerfabrik, Solingen-Ohligs; established in 1834, Klaas was a maker of fine cutlery and daggers prior to and after the National Socialist era and used a number of trademarks involving "kissing cranes" and chickens over the years.

Klein: P.E. Klein Jr.; an edged weapons maker during the National Socialist era.

Kloos: Carl Kloos, Solingen; cutlery maker during the National Socialist era.

Klüssendorf: Heinrich H. Klüssendorf, Werkzeuge, Berlin Spandau; a tool manufacturer that made various goods for the German military, such as machinegun accessories and head space gauges, during WWII.

Knaak: Deutsche Waffenfabrik Georg Knaak, Berlin; well-known firearms manufacturer and catalog retailer in business between circa 1910-1940s.

Knecht: August Knecht, Solingen; cutlery maker during the National Socialist era.

Knecht: E. Knecht & Co., Solingen; cutlery and dagger manufacturer, National Socialist era. The Knecht firm often used the abbreviation "K. & Co." below its famous "double stag" trademark.

Kneipp: Fr. Kneipp, Wiesbaden; a distributor and retailer of small arms.

Fig. 93: A stag horn handle hunting knife made by the E. Knecht Co., Solingen, stamped with the firm's famous "double stag" trademark. The word "ROSTFREI" denotes that the blade was made of stainless steel.

Knies: W. Knies, Darmstadt; manufacturer of Drillings and other sporting weapons.

Knoch: Nähmaschinenfabrik Adolf Knoch AG, Saalfeld/Saale; traditionally a sewing machine manufacturer, during WWII the firm made small arms components and head space gauges.

Knorr-Bremse: Knorr-Bremse AG, Berlin; a traditional maker of brake systems for trains, but ventured into machineguns and other small arms in the WWII period. Knorr-Bremse was still in business at the time of this writing, using a slight variant of its wartime trademark.

Knoth: Gebr. Knoth; an edged weapons maker during the National Socialist era.

Knyn: Abr. Kyn, Gräfrath; a manufacturer of daggers and other cutlery during the National Socialist period.

Kober: Wilhelm Kober & Co., Suhl; cutlery and dagger manufacturer, National Socialist era

KOBRA: Tradename used by Walter Diefke, Suhl, to market small pocket pistols.

Kodak: Kodak/Nagel-Werk, Stuttgart; a camera firm originally founded by the famed camera designer Dr. August Nagel and then acquired in 1931 by Eastman Kodak. Nagel stayed on as chief designer after the merger, but as WWII approached, the firm became more heavily involved in the production of military optics and lenses for the Wehrmacht.

Koeller: F. Koeller; Solingen-Ohligs; cutlery and dagger manufacturer, National Socialist era.

Kohl: Paul Kohl, Solingen; cutlery and dagger manufacturer, National Socialist era.

Kohl: Max Kohl AG, Physikalische Apparate u. Laboratoriumseinrichtungen, Chemnitz; a maker of precision optics, such as the ZF41, during the WWII era.

Kohlen: Friedrich v.d Kohlen; Solingen; cutlery and dagger manufacturer, National Socialist era.

Köhler: Hermann Köhler AG, Nähmaschinenfabrik, Altenburg; a maker of sewing machines, bicycles and other consumer and military goods. The firm was also a major supplier of K98 components to the Gustloff-Werke during WWII.

Köhler: Julius Köhler, Limbach; a maker of sewing machines, and other consumer and military goods. The firm was also a major supplier of K98 components to the Gustloff-Werke during WWII.

Köhler: P. Köhler, Berlin; made rifle telescopes for the military and police, circa 1920-onward. Köhler used the tradename “PECAR.” Scopes made in and around the WWII era were usually marked “4x” or “4 x 81” and were used by the military and sportsmen alike. The tradename survived WWII and was used to market scopes for sportsmen in the post-war period. Herbert Schwarz acquired the Pecar name and trademark in 1962 and was in business for many decades until only recently.

KOLIBRI: Tradename used by Franz Pfannl Waffen- und Patronenfabrik, Krems a/d Donau from about 1913-1926 to market small pocket pistols. See also *Pfannl.*

Köller: Gustav L. Köller, Solingen; maker of cutlery during the National Socialist era.

Köller: Hugo Köller, Solingen; maker of cutlery during the National Socialist era.

Kolping: H. Kolping, Solingen; maker of cutlery during the National Socialist era.

Kolum: Kolum Buswerk Eduard Becker; cutlery maker during the National Socialist period.

Kommer: Theodor Kommer Waffenfabrik, Zella-Mehlis; maker of small pocket pistols and sporting arms during the 1920-1930s.

Konig: H & A Konig, Zella-Mehlis, a manufacturer of sporting arms from about 1922-39.

Konejung: Hermann Konejung AG, Solingen; cutlery and dagger manufacturer, National Socialist era.

Korinski: Anton Ferdinand Korinski, Erlangen; made high quality Drillings and sporting arms between 1883-1920. Korinski used the “Adler Flinte” tradename.

Kornbusch: Kornbusch Waffenwerke Oberspree, Kornbusch & Company, Berlin; maker of small arms, acquired by D.W.M. in 1916, control of which later passed to Deutsche-Industrie-Werke. See also *Spreewerk.*

Kr: Not a trademark, but an abbreviation seen on barrel blanks supplied by Friedrich Krupp AG, Essen AG. See also *Krupp.*

Krahling: K. Krahling, Wetzlar; manufacturer of binoculars prior to the National Socialist era.

Fig. 94: The Oberspree Kornbusch & Co name and location roll-stamped on this WWI Gewehr 98. *Courtesy Robert Jensen.*

Krämer: Hans Krämer, Nürnberg; dress bayonet maker or marketer, National Socialist era.

Kraus: Ru. & Karl Kraus; an edged weapons maker during the National Socialist era.

Krausser: A. Krausser, Zella-Mehlis; small arms manufacturer and maker of the K. Helfricht-designed pocket pistols. Krausser manufactured at least three models of vest pocket pistols between about 1920-1930, sold under the "KH," "Helfricht" and/or "K. Helfricht" tradenames/trademarks.

Krebs: Peter Daniel Krebs, Solingen; cutlery maker during the Nazi era.

Krebs: Carl Julius Krebs, Solingen; cutlery and dagger manufacturer, National Socialist era.

Kremp: Georg Kremp, Aktiengesellschaft für Optik u. Mechanik, Wetzlar; a maker of precision optics to include microscopes, binoculars and 4x81 rifle telescopes from the late 1900s through at least the immediate post-WWII era. The firm was originally founded in 1877 by Christian Kremp, who passed it on to his son Georg in about 1920.

Krieger: Wilhelm Krieger, Solingen; cutlery maker during the National Socialist period.

Krieghoff: Heinrich Krieghoff Waffenfabrik, Suhl; manufacturer of military and sporting arms from 1929-45 in Suhl, and from 1945 onward in Ulm/Donau. The firm was founded by Ludwig Krieghoff and a partner as "Sempert und Krieghoff" in 1886 in Suhl. Originally the company concentrated solely on manufacturing sporting arms. By WWI, Herr Sempert had emigrated to the United States and his name was dropped from the company title. During the war Krieghoff made products for the military. After the close of hostilities, the firm again returned to making sporting weapons and received sizeable orders from the Netherlands which kept the company afloat during the Great Depression. During this same time, Krieghoff acquired the "Luttich" facility of Anciens Etablissements Pieper, Liege, Belgium. Later, Krieghoff managed to obtain smaller subsidiaries in Kufstein and Stertzing, Austria.[35] As National Socialist Germany edged closer to war, Krieghoff developed strong ties to the Luftwaffe. The firm became a major supplier of machineguns to the air force and also found itself producing flare guns, the MG131, MG151, P08 "Luger" pistols and the Louis Stange-designed FG42. In 1940 the firm employed 2,000 workers, and by 1944 that number had grown to 6,000. In late 1944 Krieghoff was even involved in the conversion of Carcano rifles for the Wehrmacht and Volkssturm. After WWII the Soviets dismantled the firm, destroyed the factory facilities, and sent its tooling and machines to the USSR as war reparations. The Krieghoff family fled to the Allied zone in western Germany and reestablished the company, at first in Heidenheim/Brenz, and finally in Ulm/Donau.

Kriegskorte: Kriegskorte & Co. GmbH, Stuttgart; made signal and flare pistols well before WWII; may have made firearms as early as 1940, and certainly marketed rifles after the war under the "Krico" tradename.

Krico: Tradename used by Kriegskorte & Co. GmbH, Stuttgart to market firearms. See *Kriegskorte*.

Krom: Heinrich Krom, München; cutlery and dagger manufacturer, National Socialist era.

Krombach: W. Krombach, Wetzlar; a maker of precision optics and binoculars founded in 1927. Krombach used several tradenames after WWII.

KRNKA: Refers to the designer Karl Krnka, but also used as a tradename by Österreichische Werk Anstalt, Wien to market vest small pocket automatics. See also *ÖWA*.

Krumm: Gebr. Krumm, Solingen; a maker of fine cutlery and daggers during the WWII period.

Krupp: Arter Krupp AG, Berndorf; cutlery maker during the Nazi era.

Krupp: Friedrich Krupp AG, Essen; in the context of this reference, Krupp was a manufacturer of high-quality steel for use in heavy industry, field guns and barrel making. In a larger context, Krupp controlled the lion's share of the steelmaking industry in WWII Germany, with foundries and subsidiaries in Essen, Wuppertal-Elberfeld, Mülhausen/Elsaß, Bremen-Oslebshausen, Magdeburg, Rheinhausen, Hannover, Kiel, Bremen, Breslau, and even a plant at Mariupol (Schdanow), Ukraine. Alfred Krupp managed this huge conglomerate from the early 1930s, which by 1944 had over 288,000 workers (as many as 100,000 of which were forced or slave laborers).

Krusius: Gebrüder Krusius, Solingen; maker of fine cutlery and daggers during the WWII period. Krusius used the "Gazelle" tradename to market much of its cutlery.

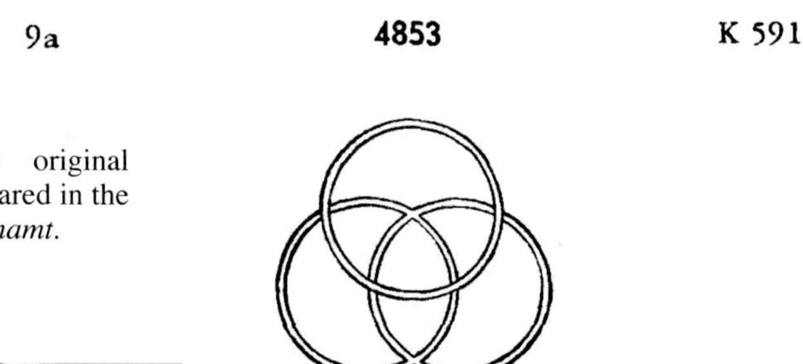

Fig. 95: Friedrich Krupp's original trademark registration as it appeared in the records of the *Deutsches Markenamt*.

Fig. 96: The Fr. Krupp, AG, Essen intertwined circle logo as seen on this WWI-era field howitzer. Krupp is usually associated with heavy industry, such as field howitzers and naval weapons, but also supplied high-quality steel to the small arms industry in the form of barrel blanks.

KRUPP-LAUFSTAHL

SPECIAL-GEWEHR-LAUF-STAHL - FRIED. KRUPP A.G. ESSEN

FRIED.-KRUPP A.G. ESSEN

NIROSTA

Fig. 97: A number of logo variants have been observed on rifle and shotgun barrels made of high-quality Krupp-produced steel.

Kugel: Gebr. Kugel & Fink; an edged weapons maker during the National Socialist era.

Kuhagen: Kuhagen, Lüchow: a post-WWII manufacturer of hunting scopes and other optics.

Kuhl: Albert Kuhl, Solingen; a manufacturer of daggers and other cutlery during the National Socialist period.

Kührt: Carl Friedrich Kührt KG, Zella-Mehlis; cutlery maker during the National Socialist period.

Küllenberg: August Küllenberg, Solingen-Gräfrath; maker of fine cutlery and daggers during the WWII period.

Küpper & Oertling: Küpper & Oertling, Solingen; manufacturer of cutlery and ceremonial daggers sold under the tradename "NEPTUN" during the National Socialist period.

KWICK: Tradename used by Carl Zander, Solingen to market cutlery.

LACRIMAE: Tradename used by Walter Diefke, Suhl, to market tear-gas pistols. See also *Diefke*.

Lang: Federn-, Draht- u. Metallwarenfabrik Rudolf Lang, Brandenburg/Havel; made various stamped sheet-metal products prior to and during WWII and was the main sub-contractor for the hasps used on binocular cases, ZF41, ZF4 and other optical containers made by other companies. Lang used a logo composed of a circle within a triangle containing a tiny letter "L."

Langenberg: J. Langenberg & Co.; an edged weapons maker during the National Socialist era that used the tradename "JULANCO."

Langenhan: Emil Langenhan, Zella-Mehlis; manufacturer of sporting arms in the 1930s.

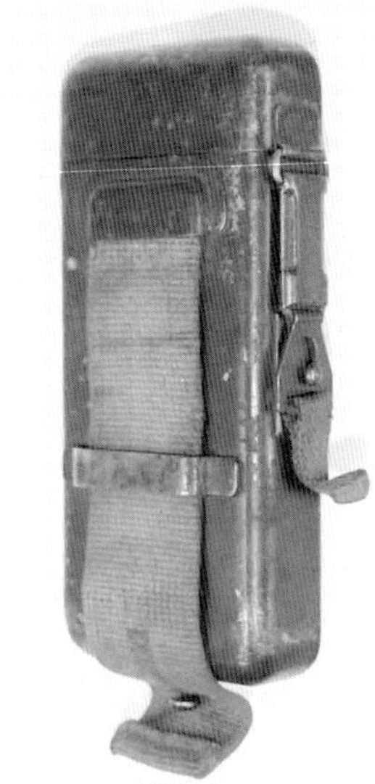

Fig. 98: A stamped sheet-metal ZF41 telescope carrying case with spring catch made by Federn-, Draht- u. Metallwarenfabrik Rudolf Lang, Brandenburg/Havel.

Fig. 99: A wooden ZF4 carrying case, again with a spring catch supplied by Federn-, Draht- u. Metallwarenfabrik Rudolf Lang, Brandenburg/Havel

Fig. 100: The trademark used by Federn-, Draht- u. Metallwarenfabrik Rudolf Lang, Brandenburg/Havel. Lang supplied the vast majority of spring catches or hasps used on nearly all of the protective cases and carriers issued with WWII German optics, such as binoculars, ZF439, ZF41, ZF4, and the MG34/42.

Langenhan: Fr. Langenhan Gewehr und Fahrradfabrik, Zella-Mehlis; small arms maker. Founded by Friedrich Langenhan from Mehlis in the mid-1800s, the company originally made flint locks, tools, knives and other goods. Eventually the firm started making pistols and received a contract in 1873 from the Sächsische Kriegsministerium to make 4,000 Modell 1873 revolvers. In 1885 Friedrich Langenhan fell ill and his son, Hermann Langenhan, took over the company and greatly expanded the facilities. The company began producing modern target and sporting rifles, self-loading pistols, flare pistols, air-guns, and even bicycles. During WWI, Langenhan produced Hebel flare pistols and 85,000 "FL-Selbstlader" pistols in 7.65mm for the military (1915-1918), and two models of vest pocket pistols for the commercial market. Hermann died in 1929 and the company was taken over by his widow and two sons, Fritz and Ernst. During the Nazi era-WWII, the company made airguns and was subcontracted to produce K98 parts and machinegun components for the Gustloff-Werke, and MP44 and other small arms components for Haenel. The firm survived WWII, but was soon incorporated into the "VEB Meteor-Werk, Zella-Mehlis."

Lanze & Fahne: See *Groten*.

Lauterjung: H. & F. Lauterjung, Solingen; maker of cutlery and daggers during the National Socialist era. Lauterjung used a number of trademarks and tradenames over the years, with "Puma" and "Tiger" the most recognizable.

Leica: Tradename of Ernst Leitz GmbH, Wetzlar; used to market cameras, binoculars and other optical products.

Leico: Tradename of Ernst Leitz GmbH, Wetzlar; used to market shooting/sighting accessories and other products.

Leitz: Ernst Leitz GmbH, Optische Werke, Wetzlar; longtime maker of cameras, optics, binoculars and telescopic rifle sights. Originally founded in 1849, by 1863 the company had passed into the control of the Leitz family. Dr. Ernst Leitz (b.1871) was an avid hunter, and thus the company introduced its first prism binocular design to the sporting market in 1907. Dr. Leitz, Ernst Leitz Jr., Ludwig Leitz and Henri Dumur made up the Board of Directors up through the 1940s. The company made precision optics for the German military in

Fig. 101: A period advertisement for binoculars made by E. Leitz, Optische Werk, Wetzlar. Leitz supplied a large percentage of binoculars, scopes and other optical devices to the Wehrmacht during WWII.

Gewehrkollimator *Leico*

DRP. Nr. 659 044.

Das neue optische Visier ist ein Richtglas für das direkte Richten von Schußwaffen, bei dem das Ziel von einem Teil des Auges ohne optische Mittel direkt gesehen wird, dessen anderer Teil eine Zielmarke (Abkommen) aufnimmt, die in der Brennebene einer Sammellinse liegt. Durch diese Sammellinse wird die Zielmarke optisch in die Entfernung „u n e n d l i c h", also praktisch in die Zielentfernung abgebildet (Kollimator).

Bild 1.

Fig. 102: A page from a WWII era brochure covering E. Leitz' "Gewehrkollimator," showing both the firm's later more familiar trademark and the cursive "Leico" tradename.

both WWI and WWII and sold its products commercially under the "Binodal," "Leica," "Leico" and other brandnames.[36]

LEONHARDT: Tradename used by H. M. Gering, Arnstadt, 1910-1920s. See *Gering*.

LEUCO: Tradename used by Johann Leupold, Bayreuth, to market cutlery and daggers.

Leue: H. Leue, Berlin; a distributor and retailer of small arms.

Leupold: Johann Leupold, Bayreuth; manufacturer of cutlery and daggers during the WWII era. Leopold used the tradename "Leuco" to market its products.

L&F: Linden & Funke KG, Iserlohn; manufacturer of various metal products, including metal slides for the WWII K98 leather sling. The company also had a subsidiary near Magdeburg.

Fig. 103: A K98 "Riemen" metal slide/adjuster made by Linden & Funke, Iserlohn.

Lignose: Aktiengesellschaft Lignose, Berlin, Abteilung Suhl; small arms maker. Lignose AG, Berlin was a major manufacturer of gunpowder and ammunition. In about 1921-1922 Lignose acquired the Theodor Bergmann, Suhl/Gaggenau firm, which was sold after the death of Theodor Bergmann (it was at this time that Schmeisser left Bergmann to start his own company and later went on to work with Haenel). Lignose AG also acquired a number of patents and continued the production of the last Bergmann vest pocket pistols marketed under the "Lignose" name. Then, in 1937 Lignose went on to purchase the August Menz company as well, and for a short time they produced a number of the former Menz pistols marked "Theodor Bergmann Erben." The company is not known to have made small arms after WWII.

Liliput: Tradename used by August Menz, Suhl to market diminutive pocket pistols. See Menz, August.

Limbacher: Limbacher Maschinenfabrik, Bach & Winter, Limbach; a maker of sewing machines, and other consumer and military goods. The firm was also a major supplier of K98 muzzle covers and K98 components to the Gustloff-Werke during WWII.

Linden: Carl Linden, Merscheid; maker of cutlery and dress daggers during the Nazi era.

Linden & Funke: Linden & Funke KG, Iserlohn. See *L&F*.

Linder: C & R Linder, Solingen-Weyer; maker of cutlery and dress daggers during the Nazi era.

Linder: C. W. Linder & Sohn, Solingen-Weyer; maker of cutlery and dress daggers during the National Socialist period. Carl Wilhelm Linder founded the knife-making company in 1870, in the small Solingen suburb of Bech. Carl Linder was born in 1869 and eventually took over the family company, moving it to a larger location in Solingen in 1903. Linder focused on military products during WWI and the cutlery export market after the close of hostilities. Carl Linder died in 1937 and the company was sold to another Solingen knife-maker, Paul Rosenkaimer. Rosenkaimer renamed the company "Carl Linder Nachf" (the German word "Nachfolger" translates as "successor"). During the National Socialist era, Linder made cutlery and dress daggers for para- and military organizations. Linder is still in business under Rosenkaimer family control at the time of this writing.

Linder: Hermann Linder Söhne, Solingen; manufacturer of cutlery and daggers during the WWII era.

Linder: Hugo Linder Deltawerk, Solingen; maker of cutlery and dress daggers during the Nazi era.

Linder: Otto Linder, Solingen; manufacturer of cutlery and daggers during the WWII era.

LINOR: Tradename used by Hugo Linder C. W. Sohn, Solingen-Weyer to market daggers and other fine cutlery during the Nazi era. See also *Linder*.

LION: Tradename used by C. Lütters & Cie Löwenwerk, Solingen to market tools and cutlery.

Little Tom: Tradename used to market pistols designed by Alois Tomiška and made by Wiener Waffenfabrik, Wien. See *Wiener Waffenfabrik*.

Loch: Loch & Hartenberger, Metallwarenfabrik, Idar-Oberstein; a manufacturer of small arms components, magazines and ammunition cartridge cases during WWII.

Loewe: Ludwig Loewe & Co., Berlin. Founded by two Jewish brothers, Isidor (1848-1910) and Ludwig (†1886) Loewe, the Loewe Co. was a large firm primarily involved in the machine tooling and small arms industries. Loewe made Gewehr 88 rifles in Berlin, owned 50% of the Belgian Fabrique Nationale d'Armes de Guerre, and in 1887, went on to acquire a controlling interest in the Waffenfabrik Mauser. However, a scandal developed over bribing of government inspectors to pass defective Gewehr 88s at the Loewe, Berlin plant during the early 1890s. Anti-Semitism was then and had been on the rise in Germany for some time. The scandal resulted in anti-Semitic accusations of treachery against the Loewe brothers personally, and Isidor decided it best to distance his company from arms making. Thus, in 1896 Loewe combined its arms manufacturing interests with the Rheinisch-Westfälische Pulverfabriken, Pulverfabrik Rottweil-Hamburg and the Deutsche Metallpatronenfabrik of Karlsruhe to form DWM (Deutsche Waffen- und Munitionsfabriken) as a separate organization.[37] Loewe itself remained an independent company, still primarily involved with machine tools, and increasingly with electrical machinery during the early 20th century. Loewe remained a major supplier of tools and machinery to the small arms industry. It was eventually merged with the "Gesellschaft für Elektrische Unternehmungen" in 1929. The firm was later known as "Loewe-Fabriken, Berlin." See also *DWM*.

Loffe: Loffe, Solingen; cutlery manufacturer during the WWII era and thereafter.

Lohmann: Lohmann-Werke AG, Fahrrad- u. Motorensättel-Fabrik, Bielefeld;
a manufacturer of bicycles and accessories. The firm made small arms components and accessories as well as other military goods and had several subsidiaries inside of Germany and in Poland during WWII.

Louper: Louper, Solingen; WWII era cutlery maker- used the "Flamme" tradename.

LOVEC: Tradename used to market pre-WWII optics made by Srb and Stys, Prague; Czechoslovakia.

Luck & Wagner: Luck & Wagner, Suhl: founded in 1879, the firm eventually specialized in sheet-metal stampings and made small arms magazine components during WWII.

Ludwig: Ernst Ludwig Optisches Werk, Weixdorf, maker of precision optics and rifle telescopes, WWII.

LUGER: Colloquialism and later became a tradename claimed and registered by Stoeger Industries. "Luger" refers to the pistol's designer, Georg Luger, and the name is used by collectors to describe all P08 pistols and the subsequent copies that were marketed by Stoeger.

LUNA: Tradename used by Cuno Büchel, Suhl to market sporting arms between about 1880-1920s.

Luna: Tradename used by A. Feist & Cie to market cutlery. See *Feist*.

Luneburg: August Luneburg, Kiel; manufacturer of sporting arms in the 1930s.

Lüneschloß: P.D. Lüneschloß, Solingen; cutlery manufacturer during the Imperial and the National Socialist periods. The firm made bayonets before and during WWI and daggers and other cutlery before and during WWII. Lüneschloß used a number of logo variations over several decades.

Lungstrass: Peter Lungstrass, Solingen-Ohligs; cutlery maker during the WWII era.

LUTKO: Tradename used by C. Lütters & Cie Löwenwerk, Solingen to market tools and cutlery.

Lütters: C. Lütters & Cie Löwenwerk, Solingen; originally founded in 1840, the Löwenwerk became well known for knives, ceremonial daggers and other cutlery between about 1900-1945. The firm also marketed goods under the "LION" and "LUTKO" tradenames.

Lüttges: E. Lüttges & Co., Solingen; cutlery maker during WWII. The firm is known to have used the "Desira" tradename.

Lutzenkirchen: Gebr. Lutzenkirchen; an edged weapons maker during the National Socialist era.

LUX: Was/is a registered trademark of Emil Lux, Remscheid; a maker of tabs and buckles as seen on K98 grenade launcher pouch straps, etc, but was also a tradename used by Carl Tillmanns Söhne, Solingen to market cutlery.

LUXOR: Tradename used by Optische Anstalt Oigee, Berlin, to market rifle telescopes prior to WWII. See also *Oigee*.

MAB: Abbreviation for Manufacture d' Armes des Bayonne, Bayonne, France. Made a number of small arms prior to the war and continued production under German occupation.

MAB is probably best known for its 7.65mm MAB Modell D pistol, based on the Browning 1910.

MAC: Abbreviation for Manufacture Nationale d'Armes de Chatellerault, Chatellerault, France. MAC made small arms and components under German occupation during WWII. See *Manufacture Nationale d'Armes de Chatellerault.*

Macero: Tradename used by Henkel & Müller, Solingen to market dress daggers and other cutlery during the Nazi era.

Madix: "Madix" Nähmaschinenteilefabrik Alfred Keller & Co. KG, Dresden; a manufacturer of sewing machine components that supplied K98 components during WWII. After the war, the firm was absorbed into the "VEB Feinwerktechnik Dresden" and made adding machines.

Mädler: Bruno Mädler, Berlin, maker of tools and precision machines, founded in 1882.

Maget: Maschinen-u Gerätebau GmbH, Berlin-Tegel; a subsidiary of Rheinmetall-Borsig, maker of machineguns, such as the MG34 and MG42, and other small arms components prior to and during WWII. The firm is also known as the "Alkett Werk II" or "Altmärkisches Kettenwerk."

Maihak: H. Maihak AG, Fabrik für techn. Meßinstrumente u. feinmech. Geräte, Hamburg; an instrument manufacturer that also made machinegun parts, accessories and FG42 magazines during WWII.

Malsch: August Malsch u. Sohn, Steinbach, Kr. Meinigen; cutlery manufacturer during the National Socialist period.

Malsch: David Malsch, Steinbach; cutlery manufacturer during the National Socialist period.

Malsch: Karl Malsch, Steinbach, Kr. Meinigen; cutlery manufacturer during the National Socialist period.

Malsch: Karl Malsch Spitzer, Steinbach, Kr. Meinigen; cutlery manufacturer during the National Socialist period.

Malsch & Ambronn: Malsch & Ambronn, Steinbach; maker of daggers and utility knives during the National Socialist era. The company was founded in 1806 by David Malsch and made various metal products throughout its history, to include cutlery. The firm did well making knives and daggers for the National Socialists and may have employed as many as 500 workers during that era. Malsch & Ambronn's facilities fell under the Soviet Zone of Occupation after WWII and its assets eventually seized by the DDR.

M.A.N: Maschinenfabrik Augsburg-Nürnberg AG; a large conglomerate, involved mainly in heavier industry: trucks, vehicles industrial products. M.A.N. did, though, make machinegun components during WWI and had corporate supervision over other steel tool makers important to the small arms industry during the National Socialist era.

Mandewirth: Ernst Mandewirth, Solingen; manufacturer of edged weapons during the National Socialist period.

Mann: Fritz Mann Feinmaschinen-, Waffen- und Werkzeugfabrik Suhl-Neundorf. Founded in 1919, Mann was responsible for the short-lived, bottle-necked "6.33mm Mann" cartridge, but had its greatest success in making vest pistols chambered in the traditional 6.35mm. In operation throughout the 1920s, Mann is reported to have gone bankrupt in 1929.

MANNLICHER: Mannlicher, Steyr, Austria; small arms designer. "Mannlicher" is a tradename, colloquialism, and a neologism all in one. It refers to Ferdinand von Mannlicher (1848-1904), perhaps the most prolific small arms designer of his era. Mannlicher worked for Wendl/ÖWG/Waffenfabrik Steyr from 1877-1904, rising to head its Research and Development section. During this period, he patented over 150 straight pull and turning bolt designs. He is probably best known for his packet-loading magazine system used on Austrian and Italian military rifles of the time. His name was subsequently used as a marketing tool, particularly in the sporting arms market. "Mannlicher" was/is also frequently used to describe any sporting rifle manufactured by Steyr.

MANNLICHER-SCHÖNAUER: Tradename and colloquialism used to refer to military and sporting rifles based on a turning bolt, receiver, and rotary-feed magazine co-developed by Ferdinand v. Mannlicher and his colleague at Steyr, Otto Schönauer.[38] The term was later claimed by Steyr-Daimler-Puch and used to market sporting arms. See also *Steyr* and *Mannlicher*.

Manurhin: Tradename used by Mre. de Machines du Haut Rhin, France. Manurhin produced Walther-designed pistols under license after WWII. See also *Walther*.

MAP: Abbreviation for Manufacture d'Armes des Pyrenees Francaises, Hendaye, France. This French arsenal began producing a Ruby-type pistol in the 1920s, marketed as the "Unique 17." MAP continued to manufacture the pistol under German occupation during WWII. The Germans subsequently redesigned the Unique by altering the rear grip and adding an external hammer. These pistols became known as the "Kriegsmodell Rr51."

Marholdt: Richard Marholdt & Sohn, Tiroler Waffenfabrik Peterlongo, Innsbruck, Austria; a longtime sporting arms retailer. The Tiroler Waffenfabrik was taken over by Richard Marholdt in 1939 and survived WWII.

MaRie: Tradename eventually used by Mayer & Riem, Arnsberg to market blank-firing pistols. In about 1933, Walter Riem, an executive at Carl Walther Waffenfabrik, Zella-Mehlis, was given permission to manufacture a blank-firing pistol of his own design. The design itself was known as the "PERFECTA," but was produced by the Germania-Werk in Zella-Mehlis until the end of WWII. After the war Mayer & Riem evolved into the UMAREX group, and in 1993, UMAREX actually bought the post-war Walther firm located in Ulm/Donau.

Fig. 104: The MARS tradename as stamped on a pre-WWII air rifle used for marksmanship training. *Courtesy Robert Jensen.*

MARS: Tradename used to market 7.65mm vest pocket pistols made by Posumavska Zbrojovka Kdyne, of Kdyne, Czechoslovakia and air pistols after the war. See *Posumavska Zbrojovka.*

MARS: Tradename used to market binoculars made or distributed by Runge & Kaulfuss, Rathenow.

MARS: Tradename used to market airguns made or distributed by Venus Waffenwerk, Oskar Will, Zella-Mehlis, circa the 1930s.

MARS: Tradename used by Weyersberg, Kirschbaum & Cie (WKC) of Solingen to market tools and cutlery.

Marx: Kr. Marx & Cie, Solingen; manufacturer of edged weapons during the National Socialist era.

MAS: Manufacture Nationale d'Armes de Saint Etienne (MAS), St. Etienne, France; small arms manufacturer, produced M.1935 7.65mm pistols and G/K43 components under German occupation during the Second World War.

Mass-Industrie: Mass-Industrie GmbH, Meßwerkzeugfabrik, Werdau; primarily a manufacturer of precision instruments that also made small arms manufacturing gauges during WWII. The firm used the acronym "MASSI" before and after WWII. After the war, the firm was revived in the DDR as the "VEB Massindustrie Werdau."

MASSI: Abbreviation for Mass-Industrie GmbH, Meßwerkzeugfabrik, Werdau. See *Mass-Industrie.*

Mauser: Alphons Mauser, Köln-Ehrenfeld; maker of barrels and containers. The son of Wilhelm Mauser, Alphons worked at the Waffenfabrik Mauser in his youth and then started his own company making welding cylinders, barrels, other types of containers, and office furniture after WWI. His firm's logo was a clear knock off of the "Mauser Barrel"- being a vertical or horizontal barrel with a flat top and bottom and the name "Mauser" running from top to bottom in the middle. This blatant trademark infringement, as well as Alphons' alleged propensity for stealing ideas from his uncle, Paul Mauser, and then claiming them as his own created a lifelong conflict between the two family members.[39]

Fig. 105: The trademark of Alfons Mauser K.G. Alfons was the nephew of Paul Mauser, and his company's trademark was a clear knock-off of his famous uncle's "barrel" trademark. *Courtesy Robert Jensen.*

Mauser: Mauser-Werke, Oberndorf a/Neckar, Württemberg: manufacturer of machineguns, rifles, pistols and other military products. Founded by the Paul and Wilhelm Mauser after they acquired the so-called Obere Werk, or Upper Works, and in 1874 acquired the Royal Württemberg Rifle Factory. Originally known as "Gebrüder W. & P. Mauser & Cie," the Mausers concentrated on the manufacture of rifles for the domestic and foreign military markets, while Paul toiled away at new designs. Wilhelm died in 1882, leaving Paul to run the company. Unfortunately, while Paul Mauser was a talented inventor, designer, and a shrewd promoter, he was not much of a business manager. Within a couple of years, the firm's financial situation worsened to the point that in 1884 the Württembergische central bank forced a reorganization of the company. Paul was left with only a minority interest in the newly-named "Waffenfabrik Mauser AG," but still retained technical and creative control, while the bank appointed its own business manager to run the financial affairs. Later, the Ludwig Loewe company in Berlin gained control of Mauser, but following a scandal over bribing of government inspectors to pass defective Gewehr 88s at the Loewe Berlin plant during the early 1890s, and resultant anti-Semitic accusations of treachery against the Loewe brothers personally, the firm wanted to distance itself from arms making. Thus, in 1896 Loewe combined its arms manufacturing interests (Waffenfabrik Mauser and 50% of Fabrique Nationale) with the Rheinisch-Westfälische Pulverfabriken, Pulverfabrik Rottweil-Hamburg, and the Deutsche Metallpatronenfabrik of Karlsruhe to form DWM (Deutsche Waffen- und Munitionsfabriken) as a separate organization. Mauser continued his work under the DWM umbrella and scored two amazing successes with the C96 pistol and G98 rifle. It was during this same period that Mauser adopted its famous "barrel" trademark, modeled after DWM's powder barrel trademark.[40] The firm concentrated on large-scale rifle contracts while Paul toiled away on self-loading pistol and rifle designs until his death in 1914. His company lived on and made many thousands of small arms for the German military during WWI. After the war, the firm struggled like most of the other German gunmakers by trying to sell firearms on the commercial market and by branching out into other products, such as taxi cabs, measuring tools and business machines. In 1922 the company's name was changed again, to "Mauser-Werke AG." Sales of firearms to police agencies and the military gradually picked up, and Mauser designers continued to work on new and innovative designs, despite the restrictions in the Treaty of Versailles. In fact, Mauser acquired a Swiss company, Mauser-Vollmer, Kreuzlingen which was used to market and export weapons and components that were actually developed and manufactured clandestinely at Oberndorf. Sales of the K98 rifle, P08, and other

Fig. 106: A Mauser pocket pistol fitted with phenolic grips. Note the early stylized initials "MW" trademark, which the firm used prior to and sporadically after the company was taken over by DWM. *Courtesy Werner Frank.*

Fig. 107: The Mauser name and location as it was roll-stamped on this WWI Gewehr 98. *Courtesy Robert Jensen.*

pistols picked up dramatically after the Nazi assumption of power. By the dawn of WWII, Waffenfabrik Mauser was providing the majority of new rifles to the Wehrmacht, pistols and other products. During the war, the company made K98s, G41(M)s, Sten copies, P38s, HSc pistols, machineguns and many other weapons for the war effort. During the same period, the company acquired or ran several subsidiaries: Werk Berlin-Borsigwalde, Werk Karlsruhe, Werk Neuwied/Rhein, and the Mauser-Werk Waldeck, bei Kassel. Waffenfabrik Mauser was occupied by French forces at the end of the war and continued to make small arms for the French until 1947. Afterwards, the majority of the tooling and machinery were divided up between the Allies as war reparations, and the factory buildings were destroyed. Mauser was resurrected a few years later, was eventually acquired by Rheinmetall AG, and continues to produce sporting arms and other products as of the time of this writing. See also *DWM*.

Fig. 108: A WWI-era English language user's manual for the Mauser Modell 1914 self-loading pistol, which incorporates the famous "barrel" trademark, adopted after the firm was taken over by DWM. *Courtesy Dr. Geoffrey Sturgess.*

Fig. 109: The Mauser trademark roll-stamped on an early "Standard-Modell." *Courtesy Robert Jensen.*

Fig. 110: An English-language advertisement from the 1940s. Mauser was not only involved in the production of small arms by this point in its history, but adding machines and precision instruments as well.

Fig. 111: An early Mauser HSc pistol, with the company'strademarketched into the slide. *Courtesy Dr. Geoffrey Sturgess.*

33 765085 M 18629

Fig. 112: Mauser's post-WWII "Original" trademark registration as it appeared in the records of the Deutsches Markenamt. Mauser had continual trouble throughout its history with other firms copying its designs and trademark, especially those in Spain and China. Mauser used the additional "Original" as one tactic to differentiate itself from those that might try to infringe or capitalize on the company's legendary "barrel."

Mauser-Vollmer: Metallwarenfabrik Kreuzlingen (MV), Switzerland; a front company purchased in 1931 by Waffenfabrik Mauser, Oberndorf a/N. Mauser-Vollmer was used by the parent firm to market and export weapons and components that were actually developed and manufactured clandestinely in Germany.[41] See also *Mauser*.

May & von Hau: May und von Hau, Solingen-Ohligs; cutlery manufacturer during the National Socialist era.

Mayer & Riem: Mayer & Riem, Arnsberg; see *MaRie*.

Mebus: Albert Mebus, Solingen; sold cutlery and daggers during the National socialist era marketed under the AMSO tradename.

Meffert: Imman. Meffert Gewehrfabrik, Berlin; founded in 1839, the firm is best known for small caliber rifles designed by Grothe, as well as shotguns. Meffert used the "Hubertus" tradename prior to market its products, but other firms, such as AKAH, used it as well. See also *HUBERTUS*.

Meinel-Scholer: C & W. Meinel-Scholer, Metallwaren, Klingenthal; manufacturer of various metal wares. The firm is best known for producing harmonicas and the LP42 flare pistol during WWII.

Meis: Carl August Meis GmbH, Solingen; cutlery manufacturer during the National Socialist era.

Meisenberg: Kuno Meisenberg, Solingen; cutlery and dagger maker during the Nazi era. The firm used the "Undine" tradename to market cutlery circa WWII.

Melcher: Arthur Melcher, Solingen-Merscheid; cutlery and dagger maker during the Nazi era. Melcher used the "ARMESO" tradename to market cutlery circa WWII.

Melzer & Feller: Melzer & Feller, Zella-Mehlis; maker of dress daggers during the Nazi era.

MENTA: Tradename used by Waffenfabrik August Menz, Suhl to market its version of the "BEHOLLA" pocket pistol between about 1915-1917. See also *Menz*.

Menz: Waffenfabrik August Menz, Suhl; maker of many types of small pistols, was one of the manufacturers of the BEHOLLA pistols- these, the firm marketed using the "MENTA" tradename (circa 1915-1917). Menz is perhaps best known for producing the diminutive 4.25mm and 6.35mm "Lilliput" pistols, from about 1920-1927. Then, when Waffenfabrik Walther, Zella-Mehlis, began marketing its highly successful PP and PPK series, Menz tried to answer with its own "Modell P.B. 'SPEZIAL,'" with "P.B." standing for "Polizei

Fig. 113: Detail from a Luftwaffe Modell-L double-barrel flare pistol, made by and marked with August Menz trademark. *Courtesy Dr. Geoffrey Sturgess.*

und Behörden." This model was not widely accepted and very few were made. Menz also made flare pistols and other military products during the Nazi era. The firm came upon hard times in the 1930s and was acquired by Lignose AG, Berlin in 1937.

Meopta: Post-WWII tradename used by Meopta Košíre, Prerov, Czechoslovakia; a maker of precision optics. After WWII, the communists nationalized much of the Czech economy, and developed collective farms and factories known as "people's enterprises." Meopta was one of these enterprises, made up from the remnants of several pre- and wartime Czech optical firms, most notably Srb & Štys Fabrik Praziser Messinstrumente, Prag (Praga), C.P. Goerz, Bratislava and the ZF4 telescope maker, Optikotechna GmbH, Prerau (Prerov).[42] A great many German units had surrendered their weapons en masse within the Czechoslovakia borders, and its optical and armaments industries had been relatively untouched by Allied bombing. To keep a segment of its people working, the Czechs in many cases simply continued assembling binoculars, scopes, small arms and ammunition from the leftover wartime components. All of this "new" material, coupled with the

Fig. 114: A Flak telescope based on the German WWII design, but made by the post-WWII Czech firm Meopta Košíre, Prerov, Czechoslovakia. *Courtesy the late William R. Smith.*

vast amounts of surrendered German material, piled up in Czech depots. With no other real exports at the time, and in desperate need of hard currency, the Czech government decided to sell its wartime surplus and newly manufactured goods on the world market, predominantly to third world countries eager to upgrade their militaries. A state trading organization called "Omnipol" was used (as actually Omnipol was already created just prior to WWII) to market, sell and export the items, as well as to purchase and sell other items on the world market. By the late 1940s, however, the surplus warehouses and depots were almost empty. When the wartime surplus finally ran out, the Czechs simply kept the factories going, manufacturing items after the war, on wartime machinery, tooling and dies. It was during this time that Meopta switched from simply assembling optics from leftover wartime components, to actually manufacturing new optics and especially, ZF4s again. Scopes destined for the commercial market were thereafter marked with the "Meopta within a double-circle" logo. Of note, Meopta was again privatized after the end of the Cold War and remains in business at the time of this writing.

Mercedes: Mercedes Büromaschinen-Werke A.G., Zella-Mehlis; a manufacturer of business machines, typewriters, et al founded in 1906. During WWII, the company made small arms components.

Mercator: Tradename used by Heinrich Kaufmann & Söhne, Solingen to market pocket knives. See also *Kaufmann.*

Merkel: Gebr. Merkel, Suhl; a longtime maker of fine Drillings and sporting firearms. Founded as a sporting arms firm in 1898, by the latter half of WWII the firm was making K98, MP44 and other small arms components. What remained of the company after the war was reorganized and taken over by the East German "people's enterprise" VEB Fahrzeug- und Jagdwaffenwerk Ernst Thälmann, Suhl.

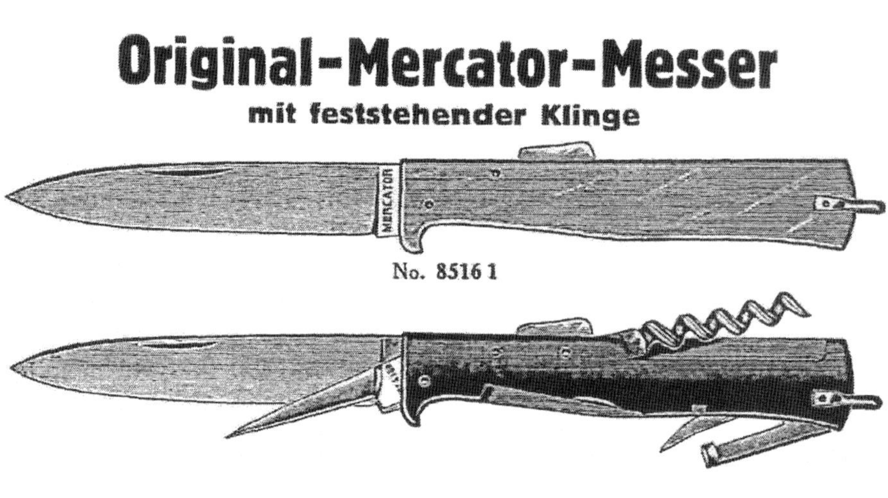

Fig. 115: An advertisement for Mercator brand knives from the 1930s made by Heinrich Kaufmann & Söhne, Solingen.

Merten: August Merten Ww, Solingen-Grafrath; cutlery and dagger maker prior to, during and after WWII.

Merz: Merz-Werke, Gebr. Merz, Metallwarenfabrik, Frankfurt-Rödelheim. Founded in 1899 by Luise Jeanette Caroline Merz and her two sons, Friedrich Karl Julius (Fritz) and Paul Georg Martin (Paul) Merz, the firm quickly became known for its innovations in the stamped sheet-metal field. Merz found initial success in the manufacture of stamped headlight housings for automobiles. During the First World War, Merz made various products for the military and eventually branched out into the pharmaceutical business (Merz Colloid-Chemische Abteilung). From 1919 onward, this department produced Novocain for the dental profession, but was sold to Bayer AG in 1929. Between the wars, Merz-Werke proper concentrated on the production of stamped office furniture and other sheet-metal products. During WWII, Merz made stamped sheet-metal scope mounts, firearms components and other stamped products for the military. The facilities were heavily damaged by Allied bombing near the end of the war. The company was up and running again by August 1945, manufacturing sheet-metal roofing. Merz was still under family control and in business at the time of this writing.

Fig. 116: A WWII MP43 fitted with a ZF41 telescope and experimental sheet-metal mount made by the Merz-Werke, Merz-Werke, Frankfurt-Rödelheim. *Courtesy the late Warren Odegaard.*

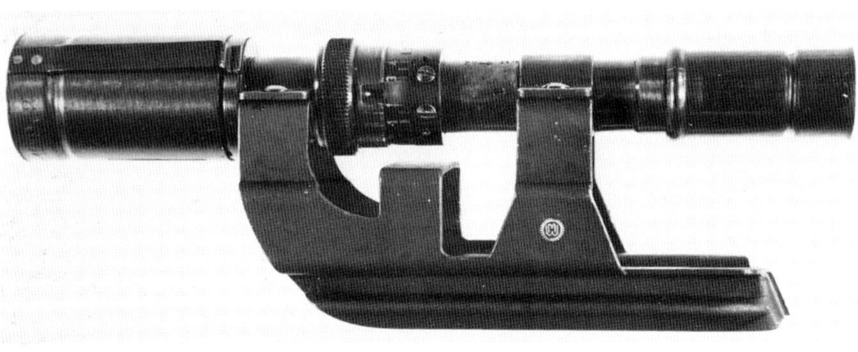

Fig. 117: Close up of the scope and mount, the latter marked with the "MW within a cogwheel" logo of the Merz-Werke, Frankfurt-Rödelheim. *Courtesy the late Warren Odegaard.*

Middeldorf: Robert Middeldorf GmbH; manufacturer of edged weapons during the National Socialist era that used the "ROMI" tradename.

Mignon: Tradename used to market rifle telescopes made by Karl Kahles, Optiker, Wein.

Miller: Miller & Val. Greiss, München; a distributor and retailer of small arms.

Mipolam: Brand name for flexible PVC, used to make green, soft, "shoot-off" muzzle caps for machine guns, the K98, G43, and a myriad of other small arms during WWII. Mipolam is a registered trademark and product of Dynamit Nobel, AG, Troisdorf.

Moeller: J. D. Moeller, Optische Werke GmbH, Wedel; a maker of binoculars and other precision optics during the Imperial and Nazi periods. The company was still in business as of the time of this writing, specializing in medical instruments.

Moritz & Gerstenberger: Moritz & Gerstenberger Waffenfabrik (EM-GE), Zella-Mehlis, manufacturer of blank weapons and air guns. Heinrich Moritz had been in business since 1916 operating under the tradename "HeiMo." He later formed a loose association with Albin Gerstenberger and they registered the "EM-GE" tradename in 1927. The firm started making air guns in 1930 marketed under the "ZENIT" tradename. Moritz & Gerstenberger also produced blank weapons and flare guns. During the latter half of WWII, the firm made MP44 and other small arms components. What remained of the company just after WWII was seized and merged into the "VEB Meteor-Werk, Zella-Mehlis."

Moritz: Heinrich Moritz Gewehrfabrik u. Schaftholzhandlung, Zella-Mehlis; a small arms maker, that was likely the predecessor of or connected to Moritz & Gerstenberger. Moritz made hunting, target and training rifles, and provided gunsmith and parts services ca. WWI-era.

Mossiers: Paul Mossiers, Breslau; a distributor and retailer of small arms.

Möve-Werke: Möve-Werke, Walter & Co. GmbH, Mühlhausen i/Thür.; See *Walter & Co.*

MSW: abbreviation for tradename used after WWII on rifle telescopes marketed by Schmidt & Bender, Biebertal / Wetzlar.

M.S.W.: A tradename and trademark used by Schmidt & Bender, Biebertal/Wetzlar to market rifle telescopes after WWII. See also *Schmidt & Bender*.
Müller: August Müller, Solingen; cutlery and dagger maker during the Nazi era. Müller used the "BISMARCK" tradename to market cutlery circa WWII.

Müller: Gottfried Müller, Herges-Vogtel; cutlery manufacturer before and during WWII.

Müller: Karl Müller; an edged weapons maker during the National Socialist era.

Müller: Paul Müller; an edged weapons maker during the National Socialist era.

Müller & Schmidt: Müller & Schmidt Pfeilringwerk, Solingen; cutlery and dagger maker prior to, during and after WWII.

Münch: Josef Münch, Brotterod; maker of dress daggers during the Nazi era.

Mundlos: Mundlos Aktien Nähmaschinenfabrik, Berlin-Magdeburg; a traditional manufacturer of sewing machines, the firm also made SG98/05 bayonets during the Imperial era and RG34 cleaning kits and K98 bayonets 1930-1940s.

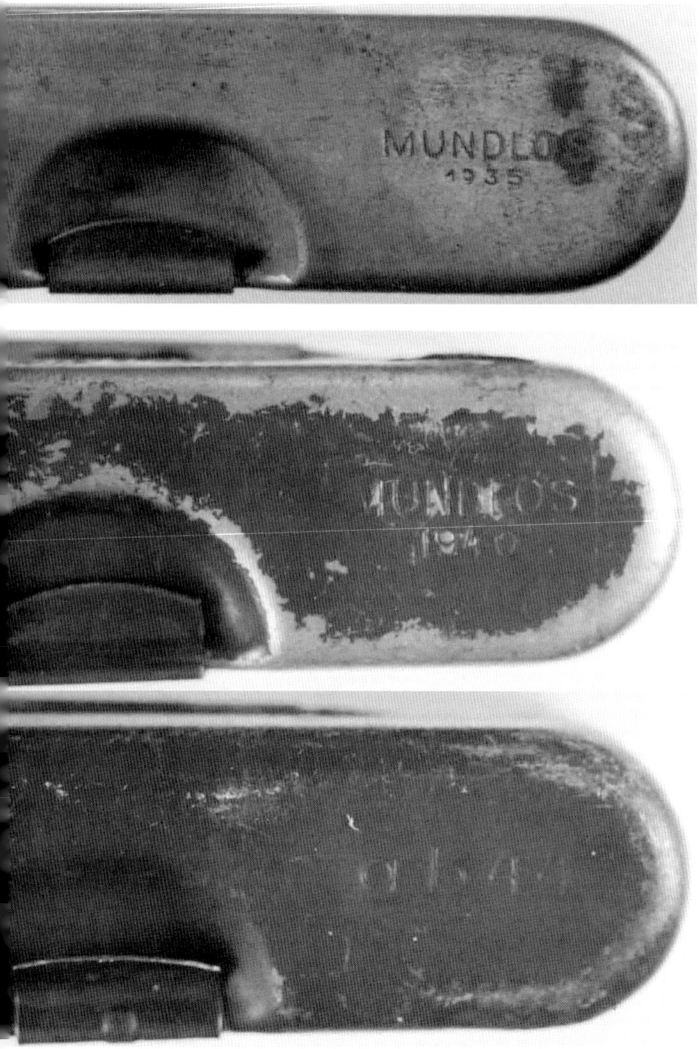

Fig. 118: The MUNDLOS (Mundlos Aktien Nähmaschinenfabrik, Berlin-Magdeburg) tradename stamped on a RG34 Cleaning Kit made in 1935 and still in use by 1940. Soon thereafter, the firm began using the military letter code "ab" instead. Near the end of WWII, MUNDLOS is thought to have been assigned yet another code, "rbl."

Nagel & Menz: Nagel & Menz, Strassburg; a distributor and retailer of small arms.

Nähmaschinenteile: "Nähmatag" Nähmaschinenteile-AG, Spezialfabriken f. Nähmaschinen- schiffchen und -teile aller Art, Radebeul, bei Dresden; a traditional manufacturer of sewing machine components founded in 1912. During WWII the firm was a principal supplier of K98 rifle components and other military goods.

NAHMATAG: Abbreviation for Nähmaschinenteile-AG. See *Nähmaschinenteile*.

NE: Abbreviation for Neunkircher Eisenwerk AG (vormals Gebr. Stumm), Homburg (Saar). See also *Neunkircher Eisenwerk*.

Neeff: Gustav Neeff, Solingen; cutlery manufacturer before and during WWII.

Neidhardt: Neidhardt & Schmidt Co., Brotterode; maker of dress daggers during the Nazi era.

NEPTUN: Tradename used by Küpper & Oertling, Solingen to market cutlery and ceremonial daggers during the National Socialist period.

Neuengamme: Metallwerke Neuengamme GmbH., Hamburg-Neuengamme; a subsidiary factory of Walther Waffenfabrik, Zella-Mehlis, operated within the confines of the Neuengamme concentration camp. The slave laborers in Neuengamme made G/K43s, pistols and components for Volksgewehre. See also *Walther*.

Neumeyer: Ernst Neumeyer, Solingen; manufacturer of edged weapons during the National Socialist period.

Neuhaus: Ferdinand Neuhaus, Solingen; cutlery manufacturer before and during WWII.

Neunkircher Eisenwerk: Neunkircher Eisenwerk AG (vormals Gebr. Stumm), Homburg (Saar); a maker of quality steel products to include barrel blanks ca. WWII.

Nickel: B. Nickel, Marburg; a post-WWII manufacturer of precision binoculars and rifle telescopes. B. Nickel worked for Zeiss, Jena during WWII. It is reported that after the war, Nickel established his own company in Marburg. B. Nickel is not thought to have made any scopes bearing the company's name before or during WWII. Of note, Nickel also acquired the assets, tradenames and Hertel & Reuss trademark after WWII.

Nico: Tradename used presumably by Nico-Pyrotechnik Hanns-Jürgen Diederichs GmbH & Co., Trittau to market gas and blank pistols.

Fig. 119: Another advertisement from the era showing DWM's "Nicorro" brand ammunition.

Nicorro: Tradename of DWM, Karlsruhe; used to market non-corrosive primed ammunition components and ammunition in the 1920-1940s. See also *DWM* and *SINOXID*.

Niedhardt: Niedhardt & Schmidt, Brotterode; manufacturer of edged weapons during the National Socialist era.

Nife: A Swedish optical firm which supplied optics to the Swedish military during WWII, included here for clarification only.

Nirosta: Trademarked type of steel used by Friedrich Krupp, Essen. See also *Krupp*.

NIROSTA: Tradename used by Gottlieb Hammesfahr Stahlwarenfabrik AG, Solingen-Foch to market cutlery. See also *Hammesfahr*.

Nitro: The word "Nitro" in either upper or lower case indicates that a firearm was made to use modern, smokeless "Nitro-Cellulose" gun powder.

Nitschke: Nitschke & Günther Optische Werke KG, Rathenow; a maker of binoculars and other optics, ca. WWI until at least the close of WWII. This firm used the tradename "GENIRA."

Noelle: Gebr. Noelle GmbH; an edged weapons maker during the National Socialist era.

Nordheim: Gotthilf von Nordheim, Zella-Mehlis, maker of the interesting 7.65mm "Deutsche Selbstlade Pistole."

Nordheim: Emil V. Nordheim, Mehlis, a maker of fine shotguns and sporting arms.

Oberspree: Waffenwerke Oberspree Kornbusch & Co., Berlin-Spandau; small arms maker. During WWI the firm made various products to include G98 rifles. Between the wars, Deutsche-Industrie-Werke, Berlin acquired a controlling interest in the company, which was subsequently renamed Metallwarenfabrik Spreewerk, Berlin-Spandau. See also *Spreewerk*.

OCULUS: "Oculus" Spezialfabrik ophthalmologischer Instrumente, Berlin. Manufacturer of precision optics and rifle telescopes during WWII, and after the war subsequently reestablished as OCULUS Optikgeräte GmbH, in the town of Wetzlar.

Ohliger: F. Ed. Ohliger, Solingen; maker of dress daggers during the Nazi era.

Ohliger: Ohliger Julius, Solingen; maker of dress daggers and other cutlery products during the Nazi era.

Oigee: Optische Anstalt Oigee, Berlin-Schöneberg; a manufacturer of fine optics and rifle telescopes 1920-1940s. Oigee used tradenames such as GNOMET, GNOM, and LUXOR to market scopes prior to WWII.

Olympia: Tradename used by various firms. Carl Walther marketed a pistol called the Olympia in the 1930s and Hermann Weihrauch marketed precision .22 caliber target rifles by the same name during the same period.

Olympia: Olympia Büromaschinenwerke AG, Erfurt; a well-known typewriter company that also made stamped metal components, loaders, grenade launcher sights, and hundreds of thousands of G/K43 rifle and other magazines during WWII.

Fig. 120: Detail from an MG13 3x observation scope made by Oigee, Berlin. *Courtesy Dr. Geoffrey Sturgess.*

Omega: Tradename used by Joseph Feist to market edged weapons during the Nazi era.

Omnipol: Omnipol, Prague (Praha); a Czechoslovakian state trading company that was created in order to market and export Czech-made firearms prior to being annexed by the Germans, and then used after WWII to sell surplus WWII German and post-war Czech manufactured arms and ammunition abroad. See also *CZ* and *Meopta*.

Orion: The word mark has been observed on G/K43 rifle receivers. It was previously thought to be a post-WWII West German small arms importer/distributor, but was registered by Hahn & Kolb GmbH & Co., Stuttgart in 1894 and re-registered by the same firm in the 1980s (a maker of small arms tools and gauges during WWII). It may also simply be a property marking of the Orion film production company. Further research is warranted.

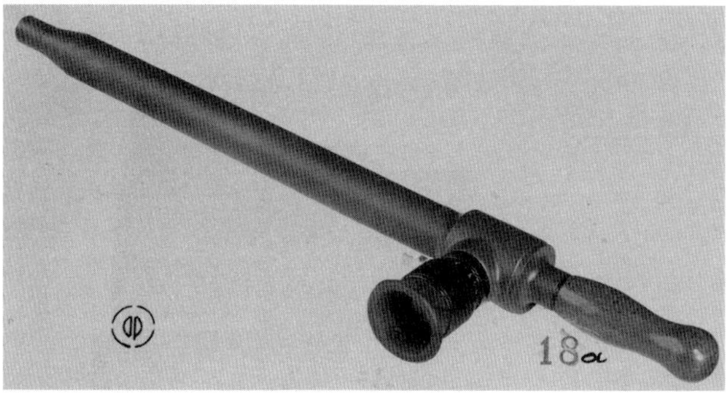

Fig. 121: Optikotechna factory photograph of a bore sight device from the wartime period showing the stylized "OP" trademark of Optikotechna, Prerau (Přerov). Optikotechna was later combined into the Meopta firm after WWII. *Courtesy Zuzana Veselá.*

Optikotechna: Optikotechna GmbH, Prerau (Přerov), Czechoslovakia; a well known maker of rifle telescopes and other precision optics, founded in 1933 by Dipl. Ing. Benes. In 1935 it became a subsidiary of the arms manufacturer Zbrojovka Brno. The company began making telescopic rifle sights in 1936. Czechoslovakia was seized and occupied by National Socialist Germany in 1938. The firm continued production under German supervision as Waffenwerke Brünn AG, Prerau and was predominantly occupied with 6x30 binocular and ZF4 telescope manufacture during WWII. After the war, Optikotechna was consolidated into the Meopta Košíre. See also *Meopta*.

Optische Präzisions-Werke: Optische Präzisions-Werke GmbH, Warsaw, Poland; made precision optics under German WWII occupation, controlled by directors appointed by Zeiss AG.[43]

Ortgies: Tradename used for pistols designed by Heinrich Ortgies, but produced by the firm Deutsche Werke AG, Erfurt. Ortgies lived many years in Liege, Belgium, and patented a handy 7.65mm pistol design in 1918. Sales were brisk, despite the post-WWI economy. Sometime around 1921, the Deutsche-Werke, Erfurt bought out Ortgies, and all of his company's tradenames, inventory, components and tooling were transferred to DW- after which Heinrich Ortgies apparently retired from the small arms business. See also *Deutsche-Werke AG.*

Oshmann: Karl Oschmann & Co., Brotterode; cutlery maker during the Nazi era.

Österreichische Waffenfabriks-Gesellschaft: Österreichische Waffenfabriks-Gesellschaft STEYR, Austria; founded by Josef Werndl, a maker of small arms and one of the companies that later was amalgamated into Steyr-Daimler-Puch. In his youth, Werndl traveled to such places as Suhl, Sömmerda, St-Etienne, Lüttich, and the United States in order to study the latest in small arms manufacturing methods. In 1855, Werndl and his brother Franz inherited their father's company, modernized the equipment, and renamed the business Waffenfabrik Josef & Franz Werndl & Co.- specializing in pistols for the Austrian government. In due course, the firm acquired the assets of Ferdinand Frühwirth and Witwe Bentz and then went public as Österreichische Waffenfabriks-Gesellschaft Steyr (abbreviated ÖWG). Werndl remained as the general manager, supervising production at the main facility and several smaller subsidiaries in and around the town of Steyr. ÖWG manufactured military goods and small arms for the Austro-Hungarian military during WWI. After the war, the company continued to develop and produce small arms in conjunction with Rheinmetall via its front company in Solothurn, Switzerland. ÖWG produced a number of firearms during the inter-war period, marketed with its ÖWG tradename. Then, in the 1920s, ÖWG branched out into business machines and automobile manufacturing, and the company name changed once again to Automobil-Fabrik Steyr-Werke AG. In 1934, Automobil-Fabrik Steyr was combined with Austro-Daimler and the Puch firms, creating the huge conglomerate Steyr-Daimler-Puch AG. For subsequent history, see *Steyr*.

Österreichische Werke Anstalt: Österreichische Werke Anstalt, Wien (ÖWA); maker of small pocket automatics chambered in 6.35mm. Created in 1919 by the Austrian government when it merged its metal-working division of its arsenal into the "Österreichischen Industriewerke," which was later cleaved off again and later reorganized as the "Österreichische Werke, Gemeinwirtschaftliche Anstalt in Wien," or "ÖWA" for short. The company was severely limited by the terms in the Treaty of Versailles, but managed to convert thousands of Italian and German rifles to other calibers. ÖWA also produced machinery, tools, motors, and a small pistol similar to the Belgian Pieper models between about 1920-1925. Unfortunately, ÖWA did not survive the Inter-Allied Military Control Commission or the post-WWI economic depression and was closed and dismantled in 1925. By 1929, all of its assets had been sold off or otherwise disposed.

Fig. 122: A small pocket pistol made by Österreichische Werke Anstalt, Wien with the company logo molded into each grip. *Courtesy the late William R. Smith.*

Othello: Tradename used by Anton Wingen to market cutlery. Also used by Henley & Co. after WWII.

Ottersbach: Ottersbach & Co., Solingen; an edged weapons maker during the National Socialist era.

Otto: Wilhelm Otto; an edged weapons maker during the National Socialist era.

ÖWA: Abbreviation/tradename for Österreichische Werke Anstalt, Wien; maker of small pocket automatics. See *Österreichische Werke Anstalt.*

P: Frequently used abbreviation/code for Polte AG, Armaturen- u. Maschinenfabriken, Magdeburg. See also *Polte*.

Fig. 123: Details of a gas pistol made by Bernhard Paatz, bearing the firm's trademark on the grip as well as the "PERPLEX" tradename. *Courtesy Dr. Geoffrey Sturgess.*

Paatz: Bernhard Paatz Waffenfabrik und Maschinenbau, Suhl, Thüringen; manufacturer of sporting arms, gas pistols, blank pistols, and small caliber training rifles. Established in 1890, the firm used a "BP" monogram/trademark and marketed single and double-barrel gas (actually the initial charge was a fine tobacco powder irritant, later changed to a crystallized tear gas substance) pistols under the "PERPLEX" tradename around 1910. Paatz also made a folding-trigger gas revolver for ladies in 1930, but is best known for its well-made .22 caliber target and training rifles produced until the first years of WWII.

Paatz: Theodor Paatz, Zella-Mehlis; a small gun making operation before WWII. This firm had given up making arms in favor of multi-spindle drill heads.

Fig. 124: Ernst Pack's distinctive logo, as etched on National Socialist era cutlery.

Pack: Ernst Pack und Söhne Stahlwarenfabrik, Solingen; maker of MG34 bolts, K98 components, bayonets, daggers and other cutlery during the National Socialist era.

Pack Ohliger & Co.: Pack Ohliger & Co., Solingen; bayonet maker during the Imperial era.

PAG: Tradname/abbreviation for Preßwerk AG, Kunstharzartikel, Essen; maker of thermoset phenolic P38 grips and G41 rifle handguards.

Parabellum: Tradename for pistols and ammunition. Parabellum is Latin for "prepare for war." The name was registered by the DWM company as early as the year 1900. The ownership of the name itself has since changed hands more than once. It was registered to Rheinmetall Waffen u. Munition GmbH, Ratingen at the time of this writing.

Pasch: Hugo Pasch Sonnl Stahlwarenfabrik; an edged weapons maker during the National Socialist era.

Pauh: Gebrüder Pauh, Solingen-Gräfrath; cutlery manufacturer noted to have made hunting/fighting knives.

Pauli: Th. Pauli Nachflg., Münster; manufacturer of Drillings and sporting weapons.

Pauls: Carl Pauls, Solingen; cutlery maker during WWII.

PDM: Not a trademark, but rather an abbreviation for the "Police Directorate Munich," or "Polizei Direktion München."

PEBRA: Abbreviation/tradename of Paul Braun Präzisionszieherei und Metallwarenfabrik, Esslingen a.N.; a supplier of high quality steel components to the small arms industry, ca. WWII.

PECAR: a tradename used by P. Köhler, Berlin to market optics in about the 1930-1940s and again after the war by Herbert Schwarz Pecar, Berlin. See also *Köhler*.

Peres: Peter Daniel Peres, Solingen; manufactured knives, swords and daggers. The firm's namesake founded the company in 1792 at age 16. By 1801 he had expanded the company into the area's first water-powered pocket knife factory, using Sheffield steel. Over the centuries the firm continued to grow and innovate, and was a well-respected manufacturer of edgeware and dress daggers by the time the National Socialists took control of Germany and through to the end of WWII.

Fig. 125: A 4x81 rifle telescope engraved with the PECAR tradename used by P. Köhler, Berlin.

PERFECTA: Tradename used by J. G. Anschütz Germania Waffenfabrik, Zella-Mehlis, Thüringen, and later Mayer & Riem, Arnsberg to market blank firing pistols designed by Walther's Walter Riem in the 1930s. The name was also reportedly used in the 1920s to market airguns made by Oskar Will, Zella-Mehlis, and has recently been used to market air guns made in China. See also *Anschütz* and *MaRie*.

PERFECTUM: Perfectum, Solingen; tradename seen on WWII cutlery and known to have been used by both Ewald Spitzer and C.D. Schaaf, Solingen.

Perimann: Perimann GmbH; an edged weapons maker during the National Socialist era.

PERLA: Tradename used by Frantisek Dusek of Opocno, Czechoslovakia, to market pistols in the 1930s.

Perlmann: Louis Perlmann; an edged weapons maker during the National Socialist era.

PERPLEX: Tradename used by Bernhard Paatz Waffenfabrik und Maschinenbau, Suhl, Thüringen, to market gas and blank pistols. See also *Paatz*.

Peterlongo: Tradename used by Tiroler Waffenfabrik, Innsbruck, Austria. See *Tiroler Waffenfabrik*.

Peters: Karl Peters, Jr.; an edged weapons maker during the National Socialist era.

Pfannl: Franz Pfannl Waffen- und Patronenfabrik, Krems a/d Donau; manufacturer of small vest pocket pistols between about 1910 to perhaps as late as the 1930s. Francois (Franz) Pfannl was a gunmaker in Krems, Austria, who was enamored with miniature firearms. Pfannl developed a number of designs of his own and miniature ammunition. Pfannl's designs were sold under the tradenames "Erika," "Kolibri," and "Kobold" (the latter chambered in 2mm Berloque caliber). Austrian small pocket pistols.

Pfeiffer: Wilhelm Pfeiffer & Co.; an edged weapons maker during the National Socialist era.

Pfeilringwerk: Müller & Schmidt Pfeilringwerk, Solingen; an edged weapons maker during the National Socialist era.

Pfestorf: Karl Pfestorf, Zella-Mehlis; a well-respected sporting arms firm founded in 1919. Pfestorf became known for 7.65mm caliber revolvers and was still in operation in Germany at the time of this writing.

Pickert: Friedrich Pickert, Zella-Mehlis; made high-quality, affordable revolvers marketed under the tradename ARMINIUS between ca. 1910-1930s. Of note, Pickert is said to have been a cousin of Carl Walther.

Pils: Franz Pils & Sohne, Steinbach; cutlery maker during the National Socialist era.

Pils: Julius Pils, Nixdorf; cutlery maker during the National Socialist era.

Pilswerk: Pilswerk Steyr, a tradename used by the Franz Pils & Sohne subsidiary in Steyr, Austria

Plücker: Friedrich Plücker, Solingen; cutlery maker during WWII.

Plümacher: Richard Plümacher, Solingen; maker of dress daggers and other cutlery products during the Nazi era.

PM: Abbreviation used by Polte-Werke, Magdeburg. See *Polte*.

Polte: Polte AG, Armaturen- u. Maschinenfabriken, Magdeburg; ammunition manufacturer 1890s-WWII. Polte was founded in 1885, and was originally known for valves and various

Fig. 126: An interesting logo from the sprawling Polte-Werke as printed on an early box of hunting ammunition.

Fig. 127: Polte advertisement from the inter-war years which made use of the firm's more familiar cogwheel logo.

heavy steel goods. The company became the largest maker of small arms ammunition for the Wehrmacht before WWII and had large subsidiaries in Duderstadt, Arnstadt, Grüneberg/Nordbahn and other smaller facilities located throughout the *Reich*.

Pose: C. Pose, Wehrausrüstungen, Berlin; a leather goods maker, particularly noted for holsters, pouches and binocular cases during WWII.

Posumavska Zbrojovka Kohout & Spol.: Posumavska Zbrojovka Kohout & Spol., Kdyne Czechoslovakia; firearm manufacturer, acquired Antonin Vilímec's company in 1935 and changed the marketing of his Colt 1908 clone from the "Slavia" to the "MARS," and then later it was sold under the tradename "DUO." Posumavska Zbrojovka Kohout & Spol. Products were frequently sold or marketed using the abbreviation "SPK."

Považská Bystrica: ZB, Považská Bystrica, Czechoslovakia; made vz.24 rifles for the Czech military and for foreign sales before WWII. When Czechoslovakia was annexed by the National Socialists, the facility was renamed Waffenwerke Brünn II, Werk Bystrica, and spent the rest of the war making K98s for the German Wehrmacht. See also *Ceskoslovenska Zbrojovka*.

PRAGA: Tradename used to market pistols made by Zbrojovka Praga, Prague, Czechoslovakia, from 1918-26. Also known as the "Praha." See *Praga Zbrojovka*.

Praga Zbrojovka: Praga Zbrojovka, Prague; an important but little-known Czech small arms firm founded in January 1918. At first, Zbrojovka Praga leased facilities in Praha-

Smichov and then established its own facilities in Praha-Vrgovice. The first design made was a 7.65mm semi-automatic pistol known as the "Praga," followed soon thereafter by the 6.35mm "Praga Model 21." In addition, the firm made Mannlicher M95 bayonets and barrels for the Austrian Schwarzlose machinegun. Over the course of its short history, Praga Zbrojovka used a "Z" displayed inside a rifle bore as its trademark. For such a small company Zbrojovka Praga at one time or another employed a veritable "who's who" of Czech small arms designers, including Karel Krnka, Vaclav Frantisek, Emanuel Holek and Frantisek Myska. With such names as these it should come as no surprise that the firm developed an innovative light machinegun which was adopted by the Czech Army in 1925 as the Praga Model 24. The firm, however, experienced financial difficulties at the same time due to an overall shortage of orders. It soon went out of business and its assets were acquired by Ceskoslovenska Zbrojovka, Brno. The Brno company not only modified and improved the Model 24 machinegun into what became famous as the zb26, but also adopted and used the Zbrojovka Praga tradenames and trademarks.

PRAHA: Tradename on pistols made by Zbrojovka Praga, Prague, Czechoslovakia, from 1918-26. Also known as "Praga." See *Praga Zbrojovka*.

Pränafawerke: Pränafawerke GmbH, Gräfrath; an edged weapons maker during the National Socialist era.

Prass: Peter Prass; an edged weapons maker during the National Socialist era.

Preßwerk: Preßwerk AG, Kunstharzartikel, Essen; "PAG" was a maker of thermoset phenolic grips and G41 rifle handguards.

Presto: Tradename used by Hagener Gußstahlwerke, Remy & Co GmbH, Hagen. See *Hagener Gußstahlwerke*.

Prym: William Prym Metallhalbfabrikate, Stolberg/Rheinland; manufacturer of snaps and fasteners seen on many WWII small arms accessories.

PS: Stylized initials for Pressstoffwerk Schöppenstedt, Paul Schnake, Schöppenstedt, thermoset plastic manufacturer, which made numerous wartime products for the military.

PTV: Not a trademark, but an abbreviation for "Polizeischule für Technik und Verkehr" or "Police Traffic and Technical School," a marking used on rifles between ca. 1928-1934 as a re-work inspection mark.

Puma: Tradename used by the Pumawerk, Lauterjung & Sohn GmbH & Co., KG, Solingen. Puma was founded in 1769 in Solingen, and made cutlery from that time through the industrial revolution and beyond. Puma made dress daggers, tableware, and popular folding and fighting knives during the National Socialist era. After WWII, the company focused on making knives for hunters and outdoorsmen.

Fig. 128: The Pumawerk logo, which in this case was stamped into the blade of a dress dagger during the Nazi era.

Fig. 129: A simplified variant marking as stamped into the blade of this Kunststoff-handled WWII-era fighting knife.

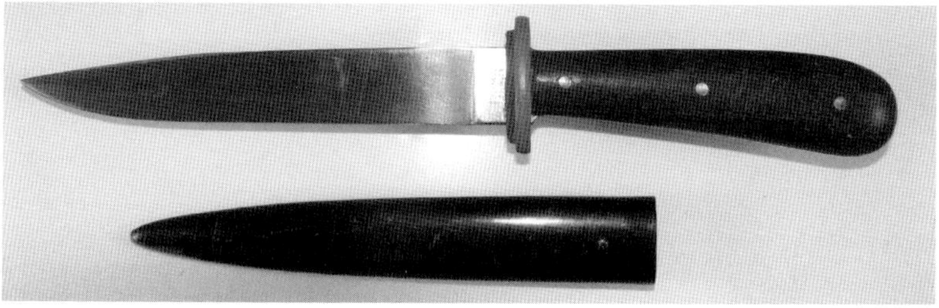

Fig. 130: Puma's post-WWII revamped and re-registered trademark as it appears in the records of the *Deutsches Markenamt*.

R: "R within shield" trademark used by RWS to market rimfire ammunition in the 1920-1950s. See also *Rheinische-Westfalische Sprengstoff Aktiengesellschaft*.

RAD: Not a tradename, but rather an abbreviation for Reichs Arbeits Dienst (German Labor Service).

Rader: Hugo Rader: Solingen; maker of cutlery during WWII period.

Radom: Not a tradename, but rather the name of a city in Poland used colloquially to describe pistols made by Fabryka Broni W. Radomiu, Radom, Poland. Piotr Wilniewczyc and Jan Skrzypiński put together the design in 1930 while working for the government armaments factory in Radom. The recoil-operated locked breech pistol was externally very similar to the Colt/Browning 1911, but internally more akin to the Browning Hi-Power.

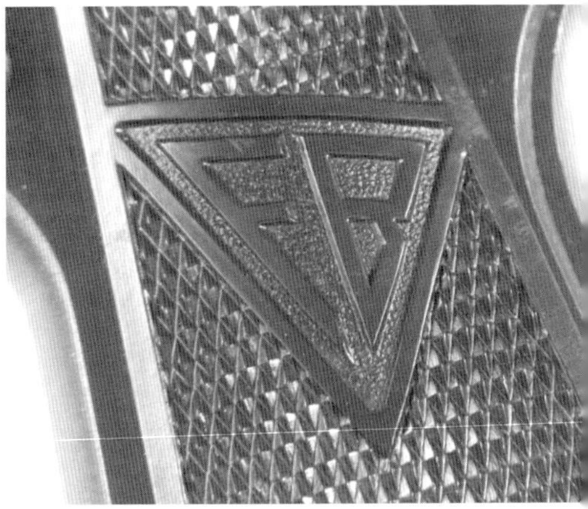

Fig. 131: A Polish P.35 9 mm pistol made from parts produced by Fabryka Broni W. Radomiu, under German occupation in Radom, Poland. This particular pistol was assembled from components in Steyr, Austria, near the end of WWII. "FB" is an abbreviation for "Fabryka Broni."

Officially adopted in 1935, the "WIS wz35" was manufactured for the Polish military from 1935-1939. The "Radom" pistol was fitted with phenolic grips with the initials "FB" molded into the grip scale on one side and "VIS" on the other. "FB" is an abbreviation for "Fabryka Broni." VIS, on the other hand, originated as "WiS"- representing the initials of the original designers (Wilniewczyc i. Skrzypiński). This was later changed to "VIS," which is Latin for "Force" or "Power." Production continued at the Radom factory under German WWII occupation, supervised by Steyr-Daimler-Puch AG; German designation "P.35(p)." Later in the war, as the Russians approached the town, tooling and parts were evacuated southward by rail and some pistols were assembled in Steyr, Austria.

Rasspe: Ernst Hugo Rasspe, Solingen; maker of sheers, garden tools, and other cutlery. Made daggers during the Nazi era sold under the REMEVE tradename.

Rau & Koch: Rau & Koch, Südwest; WWII era cutlery maker.

RD: Not a trademark, but an abbreviation seen on barrel blanks supplied by Fabryka Broni W. Radomiu, Radom, Poland. See also *Radom*.

REFORM: Tradename used by August Schüler, Suhl to market four-barrel pistols, early 1900s.

Reich: Erhardt Reich, Schweina bei Liebenstein; a cutlery maker during the Nazi era-trademark contained the words "Ein Reich Schweina."

Reichert: C. Reichert, Wien; manufactured rifle telescopes for the Austro-Hungarian Army circa WWI.

Reichswerke: "Reichswerke Herman Göring AG," a state holding company named after Hermann Göring, head of National Socialist Germany's "Four Year Plan." Founded in 1937, the concern eventually had a controlling interest in dozens of domestic and foreign firms, such as Rheinmetall-Borsig, Steyr-Daimler-Puch, and nearly every important firm in occupied Czechoslovakia, among others.[44]

Reimers: P.J.V. Antrieb Werner Reimers KG, Bad Homburg; a manufacturer of gears and clutch mechanisms, had some involvement in Volksgewehr production in the last year of WWII.

REMEVE: Tradename used by Ernst Rasspe, Solingen to market cutlery.

REMO: Tradename used by Gebrüder Rempt, Suhl to market sporting arms. See *Rempt.*

Rempt: Waffenfabrik Gebrüder Rempt, Suhl; a manufacturer of sporting arms, in business from the 1860s, was best known for WWI era flare pistols, and the "REMO" brand rifles and shotguns (similar to the "Geha"). During the latter half of WWII, the firm made MP44 and other small arms components.

Remscheid: Cuno Remscheid & Co., Auf der Höhe; maker of cutlery and daggers marketed under the "REMEVE" tradename during the Nazi era.

Remy & Co.: See *Hagener Gußstahlwerke.*

Reuleaux: Josef Reuleaux, Solingen-Wald; WWII era cutlery maker.

Reuter: Gottfried Reuter; an edged weapons maker during the National Socialist era.

RFV: Not a trademark, but rather an abbreviation for the German Finance Ministry, or "*Reichs-Finanzverwaltung.*"

Rheinisch-Westfälische Sprengstoff Aktiengesellschaft; a gunpowder and ammunition company first established in 1856. The firm used a number of trademarks over the years with its initials "RWS." RWS also eventually gained a subsidiary involved in the manufacture of thermoset plastics which was later acquired by the Nobel consortium. Renamed Alfred

Fig. 132: A carton of cartridges made by Rheinisch-Westfälische Sprengstoff Aktiengesellschaft. *Courtesy Reiner Eckert.*

Fig. 133: Another carton from Rheinisch-Westfälische Sprengstoff Aktiengesellschaft, using the firm's trademarked "U in a shield." This mark was commonly seen on RWS products, literature and advertisements up to the 1930s. *Courtesy Dr. Geoffrey Sturgess.*

Nobel und Co, Abt. Zelluloidfabrik, Werk Troisdorf, and then even later renamed Dynamit AG, Werk Troisdorf. RWS-proper continued making gunpowder and ammunition and had subsidiaries in Nürnberg, Karlsruhe-Durlach and Stadeln, bei Fürth. RWS also owned and used the "R" and SINOXID trademarks over the years.

Rheinmetall: Rheinische Metallwaren- und Maschinenfabrik GmbH. Originally founded as Rheinische Metallwaren- und Maschinenfabrik in Düsseldorf in 1889 by Heinrich Erhardt, the firm became one of the largest military conglomerates in Germany. Rheinmetall's familiar "diamond in a circle" trademark was first registered in 1894. By 1901, the company had acquired the Dreyse factory in Sömmerda and continued to produce pistols and other products using the "Dreyse" tradename. During WWI, the firm supplied the Kaiser's forces with cannon, pistols and other weapons. It was only after the war that the company became known as "Rheinmetall" and started using its familiar tradename and trademark. Though it survived the war, Rheinmetall had to lay off some 22,000 workers due to weak demand and

Fig. 134: Rheinmetall's trademark prior to the 1933 merger with the August Borsig locomotive manufacturer of Berlin-Tegal, as seen on this WWI-era field gun.

Fig. 135: One variant of Rheinmetall-Borsig's post-merger trademark. Many of the manufacturers of small arms also made other consumer goods over the years to make ends meet. This was especially true during the economic calamity and restrictions on small arms makers following WWI, and the trend continued into the 1930s. Mauser, Walther and Rheinmetall-Borsig all made business machines at one time or another which were marketed using their already familiar trademarks and logos. *Courtesy the late William R. Smith.*

Fig. 136: The company trademark as marked on this rare WWII Luftwaffe FG42 rifle. *Courtesy Dr. Geoffrey Sturgess.*

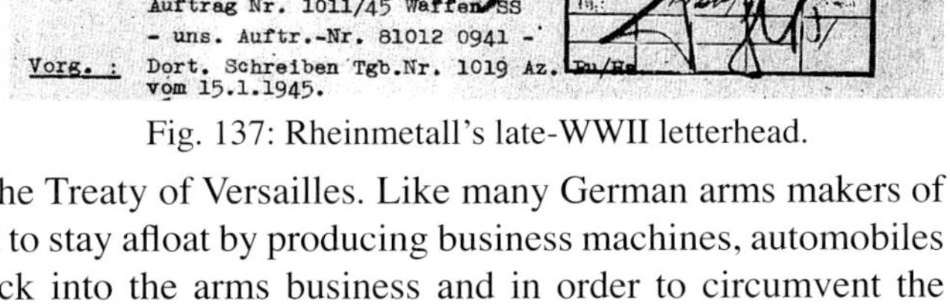

RHEINMETALL-BORSIG AKTIENGESELLSCHAFT

An den
Reichsführer SS
Befehlshaber des Ersatzheeres
Deutscher Volkssturm
(1) Berlin - Grunewald
Douglasstrasse 7

3- FEB 1945

Vo. Tgb. Nr.

Betreff : Volksgewehr.
Auftrag Nr. 1011/45 Waffen SS
- uns. Auftr.-Nr. 81012 0941 -

Vorg. : Dort. Schreiben Tgb.Nr. 1019 Az.
vom 15.1.1945.

Fig. 137: Rheinmetall's late-WWII letterhead.

the severely limiting terms in the Treaty of Versailles. Like many German arms makers of the time, Rheinmetall managed to stay afloat by producing business machines, automobiles and other products. To get back into the arms business and in order to circumvent the treaty and market its products abroad, in 1929 the company acquired a 90% ownership of Waffenfabrik Solothurn AG in Switzerland. In 1933, Rheinmetall went on to acquire the locomotive factory of August Borsig in Berlin-Tegel, thus becoming known as Rheinmetall-Borsig. Facilities were also acquired in Berlin-Marienfelde, Apolda, Guben, Unterlüß, and Liege, Belgium, in short order. In addition, Rheinmetall-Borsig eventually gained a controlling interest in the Maget firm and had a significant supervisory role at Steyr-Daimler-Puch, Steyr, Austria.[45] Rheinmetall-Borsig's chief small arms designer during this era was Louis Stange, who developed the FG42 and several innovative late-war weapons. The conglomerate made thousands of different products and components for the Wehrmacht during WWII; most notably artillery, machineguns, the FG42, MG34, MG42, MP44 curved barrels, Volksgewehre and other small arms. Rheinmetall survived WWII and is still a major defense contractor with world-wide sales.

Rh.S.: Abbreviation for Rheinmetall, Sömmerda. See *Rheinmetall.*

Ritter: Kuno Ritter, Solingen; cutlery manufacturer during the Nazi era.

RITTER: Tradename used by Anton Wingen Jr., Solingen to market cutlery and tools.

RJ: Not a trademark, but rather an abbreviation for the German Justice Ministry, or " Reichs-Justizministerium."

Röchling: Röchling´sche Eisen- u. Stahlwerke GmbH, Stabfedernwerk (Coenders), Wetzlar; a producer of high-quality steel metal that supplied barrel blanks in the Imperial era and other products to the German military. Röchling continued to supply the Wehrmacht with steel products before and during WWII. In addition, the firm had a close relationship with August Cönders, a busy consultant who was primarily involved with the development of artillery pieces. Near the end of WWII, Röchling and Cönders submitted two innovative designs for Volksgewehre that were not put into production. The prototypes of these weapons were found in the "Small Arms Section of the Hessische Industrie-Werke in Wetzlar."[46] Just what connection the Röchling, Wetzlar, and Hessische Industrie-Werke, Wetzlar firms had is not exactly known; however, the author is certain that some sort of subsidiary arrangement existed, or that the two firms were actually one and the same.

Rodenstock: G. Rodenstock Optische Werke, München; a maker of precision optics, binoculars and rifle telescopes. The company was founded in 1877 by Josef Rodenstock, and presumably used the initial "G" to honor Josef's father, Georg. In the early years, the company concentrated on producing spectacles and measuring instruments. In 1883, the firm was moved to Munich and set about making camera lenses, telescopes, binoculars and other precision optics. Rodenstock used a number of tradenames over the years, such as "Darod," "Elfa," "Euclar," "Geromar," "Lumar," "Robra," "Thepris," Tearnar," and others. The firm also made thousands of examples of the Fernglas 08 and other optics for the German military during WWI and continued to produce optics for the WWII Wehrmacht.

Röhm: Röhm-Gesellschaft, Präzisionswerkzeug-Fabrik, Zella-Mehlis; manufacturer of tooling, and later firearms and accoutrements. Founded in 1909 by Heinrich Röhm at age 27, the company originally focused on precision cutting tools and drill bits. After WWI the company expanded and grew, but it is not known if it became involved in weapons production prior to 1945. Just after the war it is known that many machines and skilled workers from the Walther plant were transferred to the Röhm facility to make copies of Walther designs for the Eastern Bloc; eventually the company facilities were nationalized and seized by the DDR. The Röhm company itself was reestablished in Sontheim/Brenz

Fig. 138: A Waffen-SS soldier scans the battlefield in this WWII propaganda photograph. His binoculars appear to be marked with the G. Rodenstock Optische Werke trademark.

after WWII, whereupon Röhm diversified the product line to include air guns, small-caliber revolvers, blank and gas pistols, and accessories.

ROJA: Tradename used by Emil Busch AG, Rathenow to market optical products.

Römer: Ernst Römer, Solingen; cutlery manufacturer during the Nazi era.

Römerwerk: Römerwerk A.G, Suhl; smalls arms manufacturer, also involved in making various small arms components, such as barrels and air rifles. Römerwerk had close ties to the Chr. Funk firm and went bankrupt in 1931.

ROMI: Tradename used by Robert Middeldorf GmbH to market edged weapons during the National Socialist era. See *Middeldorf.*

Romüso: Romüso, Merscheid; an edged weapons maker during the National Socialist era.

ROSTFREI: not a brandname, but rather a German-language designation for "stainless steel"; frequently seen on WWI-WWII era German cutlery and other military products.

Rother: August Rother, Solingen; cutlery and dagger manufacturer during the Nazi era.

Röttgen: Ernst Röttgen, Solingen; WWII-era cutlery maker.

Rottweil: Pulverfabrik Rottweil-Hamburg, Rottweil and Hamburg; a maker of gunpowder and ammunition, known especially for its sporting cartridges and shotgun shells. The powder factory at Rottweil /Neckar goes back at least to the 15th century. Pulverfabrik Rottweil was merged with Rheinisch-Westfälische Pulverfabriken and the Deutschen Metallpatronenfabrik of Karlsruhe to form DWM (Deutsche Waffen- und Munitionsfabriken) in 1896. The name and later tradename "Rottweil" only really became famous in the 1920s. During WWII the company supplied powder and ammunition to the Wehrmacht. After the war, the company was acquired by the RUAG group based in Switzerland. See also *DWM.*

Ru: Abbreviation for Ruhrstahl AG, which was marked on barrel blanks and other products made by the firm ca. WWII. See also *Ruhrstahl.*

Fig. 139: The familiar logo of Pulverfabrik Rottweil-Hamburg. *Courtesy Bob Cutler.*

Ruberg: Ruberg & Renner, Inh. Jos. Ruberg, Hagen; a manufacturer of bicycle components founded in 1918 by Felix Ruberg. The firm was making inexpensive cameras marketed under the "Dixi" tradename by about 1930. Once the war began, Ruberg focused on making small arms components and machinegun ammunition carriers and magazines. The company survived WWII but ceased operations in 1953.

RUBY: A colloquialism/description used to describe a standard pattern of Spanish-French WWI- era pistols. The "Ruby" pistol design was a simplified copy of the Browning 1903, chambered in 7.65mm. In 1915 the French government contracted the Spanish company Gabilondo y Cia Urresti to deliver 30,000 pistols per month. The firm did not have the capacity to produce that amount, and thus subcontracted much of the order to other firms. By 1918 there were 39 separate Spanish companies making "Ruby" pistols or components, and nearly a million pistols are thought to have been made during this time. When France capitulated to the Germans in 1940, tens of thousands of the "Ruby" were captured and used by the Wehrmacht and fascist para-military auxiliaries up until the end of WWII.

Rüdiger: Fa. RÜDIGER, Hildesheim; made rifle telescopes for military and sporting arms, circa WWI-WWII.

Ruf: (Schütz) Ruf & Co, Kassel; a maker of precision optics. Ruf merged with the Carl Schütz Optische Werke A.-G., Kassel ca. WWI. The company became known as "Schütz, Ruf & Co, Kassel" and continued to make binoculars and rifle scopes for the military through at least the end of WWII. See also *Schütz*.

Ruhnke: Fa. Carl Ruhnke, Berlin; a maker of precision optics, binoculars and rifle telescopes. Founded in 1896 by Carl Ruhnke (1874-1922), the company made optics through and after WWII and is known to have marketed optics under the "Televist" tradename.

RUKA: Tradename/abbreviation for Runge & Kaulfuss, Rathenow, See *Runge & Kaulfuss*.

Fig. 140: 6x24 power binoculars made by Runge & Kaulfuss, Rathenow and marked with both the "RUKA" and "MARS" tradenames.

Runge & Kaulfuss: Runge & Kaulfuss, Rathenow; a maker of precision optics, binoculars and rifle telescopes up through at least the end of WWII. This company is known to have marketed optics under the "HELLA," "MARS," and "RUKA" tradenames.

Ruhrstahl: Ruhrstahl AG, Preßwerk, Bielefeld-Brackwede; made a small number of VG1 Volksgewehr at the end of WWII.

Ruhrstahl: Ruhrstahl AG, Abt. Annener Gußstahlwerk, Witten-Annen/Ruhr: maker of high-quality steel barrel blanks for machineguns and K98s used by numerous small arms manufacturers during WWII. The firm frequently marked its products with the abbreviation "Ru."

Rusche: W.O. Rusche, Merscheid; an edged weapons maker during the National Socialist era.

RWS: Abbreviation for Rheinisch-Westfälische Sprengstoff Aktiengesellschaft. See *Rheinische-Westfalische Sprengstoff.*

SA: Not a tradename, but an abbreviation for the NSDAP "Sturmabteilung", a.k.a Brown Shirts or Storm Troopers

Sackreuter: L. Sackreuter, Frankfurt a.M.; a distributor and retailer of small arms.

SACM: Société Alsacienne de Construction Mécanique (SACM), Chalet; a French state arsenal that produced the M1935A pistol during WWII under German occupation.

Sächsische Gußstahlwerke: Sächsische Gußstahlwerke Döhlen AG, Freital; made barrel blanks for small arms, such as the K98 and G43, during the 1930-1940s. Products from this firm were often marked "Dö."

Fig. 141: The letterhead of Sächsische Gußstahlwerke Döhlen AG, Freital. The firm used the letter "D" on a stylized mountain top as its logo. *Courtesy Jon Speed.*

Sasse: Dietrich Sasse´s Söhne, Wien; a maker of K98 barrels during WWII.

Sauer: J.P. Sauer & Sohn Gewehrfabrik, Suhl; sporting and military small arms maker. Founded in 1751, the factory originally produced high-quality black powder weapons, and by the late 1800s was a well-respected manufacturer of sporting arms. In the early 1880s the firm temporarily merged with Spangenberg to form the Spangenberg & Sauer Royal Weapons Factory, made needle rifles and single shot pistols, and by the latter half of the century was producing Reichsrevolvers. By 1885 the association ended. Renamed J.P. Sauer & Sohn Gewehrfabrik, the company went on to manufacture revolvers for the military, and then in about 1900 began commercially marketing the "Bär" pistol. Later, Sauer became involved with Georg Roth and produced its first self-loading pistol, the Roth-Sauer. By 1913 the firm was producing its own pistol design. Sauer went on to make the G98 during WWI, "Fortuna" typewriters and sporting arms between, and the K98, MP44, VG1, 38H, other small arms, and even ceremonial daggers during the Nazi era through the end of WWII. After the war, the Sauer facilities in Suhl fell under Soviet occupation and the family fled westward with most of the company's patents and production documents and drawings. Re-founded in Eckernförde, Sauer began producing sporting weapons again and then licensed some its designs to Hämmerli in Lenzburg, Switzerland. Eventually, Sauer was merged with the Schweizerische Industrieanlagen Gesellschaft, or S.I.G., and became known as SIG-Sauer. See also *S.I.G.*

Fig. 142: The Sauer name roll-stamped on a WWI Gewehr 98. *Courtesy Robert Jensen.*

Fig. 143: J. P. Sauer & Sohn, Suhl sales brochure from the 1920s. Sauer used a number of trademarks over the years, to include the one shown here as well as more than one stylized variation of its initials, "S & S."

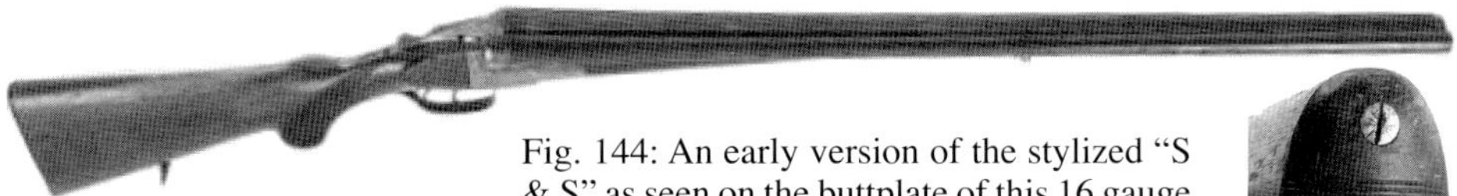

Fig. 144: An early version of the stylized "S & S" as seen on the buttplate of this 16 gauge shotgun.

Fig. 145: A late-War Modell 38H made by J. P. Sauer & Sohn, Suhl with the firm's later more familiar trademark molded directly into the phenolic grip plates.

Fig 146: A photograph taken by the U.S. Army just after the Sauer facility was captured by Allied troops. Note that the guard has obtained a factory assembly plaque with all of the actual components needed to assemble the 38H pistol. *US NARA*.

Fig 147: A post-WWII Sauer advertisement after the company was reestablished at Eckernförde in the Western Zone. Note the use of the "Original Sauer" trademark as illustrated at the beginning of this text.

Fig. 148: A K98 "Reimen" metal slide/adjuster made by Schmöle & Co., Metallwerke, Menden.

SC: R. & G. Schmöle & Co., Metallwerke, Menden; a company that manufactured ammunition components and also made metal slides/adjusters for the WWII K98 leather sling. The company was founded in 1853 and Gustav Adolf Schmöle later took over the management and expanded the company. In 1891 it had 130 employees and 1,300 by 1943. Schmöle's logo included a pine tree between the letters "S" and "C."

Schaaff: E.D. Schaaff, Solingen; cutlery manufacturer during the Nazi era. Schaaf used a trademark which consisted of a fisherman with a pipe and also marketed edged weapons under the "Perfectum" tradename.

Scharfenberg: Georg Scharfenberg, Regensburg; made interesting single-shot target pistols which were distributed by Josef Häring in Regensburg, circa 1910.

Scheidt: Eugen Scheidt, Solingen-Ohligs; cutlery and dagger manufacturer during the Nazi era.

Scheintod: Tradename used by H. Burgsmüller & Söhne GmbH to market an interesting three-barrel "pot metal" gas and blank pistol. The German word "Scheintod" translates as roughly "catalepsy," referring to the effect the firing of this pistol would have on individuals due to the intensity of flash and report. These lightweight pistols will have grips marked "Scheintod Repetier Pistole" and appear similar to the two-barrel PERPLEX brand made by Bernhard Paatz Waffenfabrik und Maschinenbau, Suhl, Thüringen.

Schichau: Schichau, F., GmbH, Schiffswerft, Maschinen- u. Lokomotivfabrik, Elbing; traditionally a manufacturer of locomotives and large ships. This firm was founded by Ferdinand Schichau in 1837 and made all kinds of vessels for the Imperial and National Socialist navies. Near the end of WWII, Schichau was also involved in producing Volksgewehre.

Schilling: Valentin Christoph Schilling, Suhl; manufacturer of military rifles, revolvers, Bergmann pistols and sporting arms. The firm made bayonets and other products for the military before and during WWI. Schilling was also part of the "Suhl Consortium" which collaborated in assembling rifles. Products made by the Schilling firm were frequently marked with the founder's initials, "V.C.S."

Fig. 149: The Schilling name and location roll-stamped on this WWI Gewehr 98. *Courtesy Robert Jensen.*

Schlegelmilch: Hermann Schlegelmilch, Zella-Mehlis; a manufacturer of barrels, primarily for commercial sporting arms. Schlegelmilch marked their products with the initials "H.S." within an oval.

Schlessinger: Peter Schlesinger, Metallwarenfabrik, Offenbach/Main, a maker of RG34 cleaning kits.

Schlieper: Carl Schlieper, Solingen; WWII era cutlery maker.

Schlimbach: Josef Schlimbach, Solingen; WWII era cutlery maker.

Schmeisser: Tradename used by C.G. Haenel, Suhl to market pocket pistols designed by its chief designer, the famous Hugo Schmeisser (1884-1953). Earlier, Schmeisser had worked with his father, Louis Schmeisser, at the Bergmann firm in Gaggenau. There, Schmeisser was responsible for developing the world's first practical sub-machinegun, the MP18/I. After WWI, Hugo and his brother Hans formed their own company, Industriewerk Auhammer Koch & Co., Suhl. Hugo Schmeisser developed a number of SMG and pistol designs and protected them with patents under his own name. When the company came upon hard financial times, the Industriewerk was merged with C.G. Haenel, Suhl. Schmeisser accepted a position within Haenel and continued to develop weapons such as the MP28. His name is most often associated with the MP38/MP40 series of WWII-era sub-machineguns, chiefly because the weapon used a magazine patented by Schmeisser. His greatest achievement was in the development of the MP44/StG44, fielded near the end of WWII. Schmeisser was kept in Soviet custody after the war at Izhevsk, where he helped develop weapons for the Red Army. He was finally returned to East Germany in 1952 and died in 1953. See also *Haenel.*

Fig. 150: A Modell II pocket pistol designed and patented by Hugo Schmeisser and manufactured at C.G.Haenel, Suhl. *Courtesy Ed Buffaloe.*

Schmidt & Bender: Schmidt & Bender, Biebertal /Wetzlar; a post-WWII manufacturer of scopes and other optics.

Schmidt & Habermann: E. Schmidt & Haberman Jägdgewehrfabrik, Suhl, maker of sporting arms marketed under the "Escha" tradename, circa 1920-40. During the latter half of WWII the firm made MP44 and other small arms components.

Schmidt: Carl Schmidt, Hamburg; a distributor and retailer of small arms.

Schmidt: A.G. Schmidt & Sohn; a manufacturer of daggers and other cutlery during the National Socialist period.

Schmidt: Carl Schmidt Sohn AG, Solingen; maker of tools and cutlery. Founded in 1829 by Carl Schmidt, the company is known for its high quality cutlery and National Socialist era dress daggers.

Schmidt: Eugene Schmidt, Solingen-Ohligs; maker of cutlery and fine daggers during the Nazi era.

Schmidt: Herbert Schmidt Waffenfabrik, Zella-Mehlis; small arms maker founded in 1919. Originally a small "mom and pop" workshop that made shotguns and Flobert rifles, the company grew and made products mostly for South American countries, but also received orders from countries such as Sweden. During WWII the firm was kept busy manufacturing small parts and repairing K98 rifles. The company was reestablished after the war in Ostheim (Rhön).

Schmidt: H.W. Schmidt, Metallwaren, Döbeln; a tool manufacturer that supplied K98 buttplates and other small arms components during WWII.

Schmidt: J. A. Schmidt & Sohn, Solingen; cutlery and dagger manufacturer during the Nazi era.

Schmidt: Julius Schmidt und Söhne, Riemberg, area Breslau, Silesia, maker of fine cutlery, K98 bayonets.

Schmidt: Rudolf Schmidt, Solingen; WWII-era cutlery manufacturer.

Schmidt & Bender: Schmidt & Bender, Biebertal /Wetzlar; a post-WWII manufacturer of scopes and other optics founded in 1957.

Schmitz: Friedrich August Schmitz, Solingen; maker of cutlery and fine daggers during the Nazi era.

Schmöle: R. & G. Schmöle & Co., Metallwerke, Menden. See *S&C*.

Schneider: Gustav Schneider Nachf.; an edged weapons maker during the National Socialist era.

Schneider: Hermann Schneider, Solingen-Auf der Höhe; maker of cutlery and daggers during the Nazi era.

Schneider: Hugo Schneider, AG. See *HASAG*.

Schneider: Schneider & Co. KG, Optische Werke, Werk Göttingen and Werk Bad Kreuznach, maker of cameras and rifle telescopes, such as the 4-power Dialytan and other optical products.

Fig. 151: A 1940s advertisement from Schneider & Co. KG, Optische Werke, which had facilities in both Göttingen and Bad Kreuznach, showing both trademark variants.

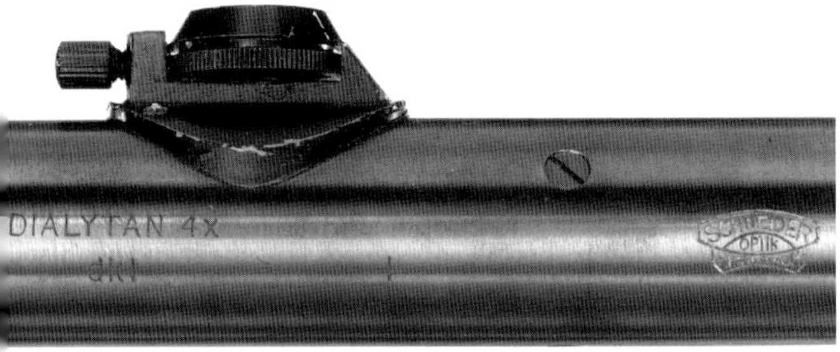

Fig. 152: Detail of an interesting transitional 4x DIALYTAN scope, bearing both Schneider's traditional trademark and its wartime ordnance code, "dkl." *Courtesy Dr. Geoffrey Sturgess.*

Schnittert: Schnittert Abr, WASSO, Solingen-Wald; cutlery and dagger manufacturer during the Nazi era. Snittert used the tradename WASSO to market many of its products.

Schott: Schott & Gen., Jena: optical company in the WWII era.

Schrick: Emil Schrick & Sohn, Solingen; cutlery and dagger manufacturer during National Socialist era.

Schricker: Schricker & Co., Fürth-Vach, bei Nürnberg; made a small number of VG1 Volksgewehr at the end of WWII.

Schüler: August Schüler, Suhl; small arms maker, best known for its "Reform Pistols" first patented in 1906 (DRP 177023). Production of the Reform series began about 1907-1910, but the firm also made other products, such as gas pistols, blank pistols and Aydt-type Schützen rifles- frequently marked or otherwise branded with the firm's abbreviation, "ASS."

Schurk: Adam Schurk, München; maker of Schützen target rifles.

Schüttelhöfer: Arthur Schüttelhöfer & Co., Solingen; cutlery and dagger manufacturer during National Socialist era. The firm frequently marketed daggers under the "ASSO" tradename.

Schütz: Carl Schütz Optische Werke A.-G., Kassel; maker of high quality optical devices for the commercial and military markets. Schütz marketed 10x binoculars marketed under the URANOS brandname ca. WWI. The company was later taken over by the Ruf & Co. firm, and the Schütz/Ruf company, a.k.a. "Schütz, Ruf & Co, Kassel," made binoculars and rifle scopes for the military through at least the end of WWII. See also *Ruf.*

Schwarzlose: Andreas W. Schwarzlose GmbH, Berlin; a company named for the well-known designer of innovative self-loading pistols (including the blow-forward Model 1908), 1907/12 machinegun and other small arms. Schwarzlose also reportedly had a facility in Linsenplatz. The primary Berlin factory was closed down in 1919, though the company was reportedly in business until about 1927. Once this was closed, Schwarzlose himself worked for other firms until his death in 1936.

Fig. 153: The legendary Schwarzlose 1907/12 machinegun, the silhouette of which was later used as Andreas W. Schwarzlose's company trademark.

Seibel: W. Seibel, Mettmann; an edged weapons maker during the National Socialist era.

Seibert: W&H Seibert, Wetzlar; manufactured rifle telescopes marketed under the "Jaguar" tradename, circa WWI-1920s.

Seidenweberer Berga: Seidenweberer Berga C.W.Crous & Co., Berga/Elster, maker of precision optics and rifle telescopes during WWII.

Seilheimer: Paul Seilheimer, Solingen; cutlery and dagger manufacturer during National Socialist era.

Sellier & Bellot: Sellier & Bellot Patronen-, Zündhütchen- u. Metallwarenfabrik AG, Prag, an ammunition manufacturer/retailer.

Sempert & Krieghoff: Sempert & Krieghoff, Suhl, firearms maker and retailer, forerunner of Heinrich Krieghoff Waffenfabrik, Suhl. See *Krieghoff*.

Senta: Tradename used by Hermann Linder Söhne, Solingen to market cutlery and daggers during the WWII era.

Servatius: Hugo Servatius, Solingen; maker of cutlery and fine daggers during the Nazi era.

Siebold: Siebold, Kassel; made high-quality Schützen target rifles around the year 1900.

S.I.G.: Schweizerische Industrie Gesellschaft S.I.G., Neuhausen, Switzerland, maker of both military and sporting arms, established 1857. S.I.G. merged with J.P. Sauer long after WWII to become SIG-Sauer.

Siemens & Halske: Siemens & Halske AG, Berlin; a large electrical engineering conglomerate founded in 1897 when the former *Telegraphen-Bauanstalt Siemens & Halske* became an openly-traded stock company. Primarily involved in the manufacture of electrical goods, the firm did, however, make rifle and MG receivers, and is thought to have made barrel blanks for the German military ca. WWI as well. Siemens & Halske later evolved into what is now known as Siemens AG.

Siemens-Werke: Friedrich Siemens-Werke AG, Unternehmung für Wärmetechnik und Gasapparatebau, Wien; another subsidiary of the Siemens conglomerate that made ammunition cans and other items for the Wehrmacht during WWII.

Simon: Otto Simon, Steinbach; maker of cutlery and fine daggers during the Nazi era.

Simplex: Model of Bergmann pistol that was manufactured by several different firms.

Simson: Simson & Co., Suhl, also known as Waffenfabrik Simson, Suhl. Founded in 1856 by two Jewish brothers, Löb and Moses Simson, the company initially focused on manufacturing carbon steel axes, chisels, pipe, and gun barrels. Manufacture of firearms followed shortly thereafter. The firm made Nagant military service revolvers for Argentina during the 1890s. Over the years the company remained in family control, and in time branched out into bicycles and automobiles. During WWI, Simson supplied the German armed forces with aircraft engines, small arms and machineguns. After the war, Simson was the only company allowed by the Inter-Allied Control Commission to supply the Reichswehr with machine guns, rifles and P08 pistols from 1923-1933. The machines, fixtures and tooling for P08 production were obtained from the defunct Erfurt government arsenal. In 1922, the firm also began marketing its own 6.35mm pocket-pistol. In about 1934, the Nazis seized control of the company from the brothers Arthur and Julius Simson and the name of the company was changed to Berlin-Suhler Waffen- und Fahrzeugwerke GmbH. The Simson brothers endured many legal entanglements after the Nazis seized

Fig. 154: The Simson company name roll-stamped on a WWI Gewehr 98. *Courtesy Robert Jensen.*

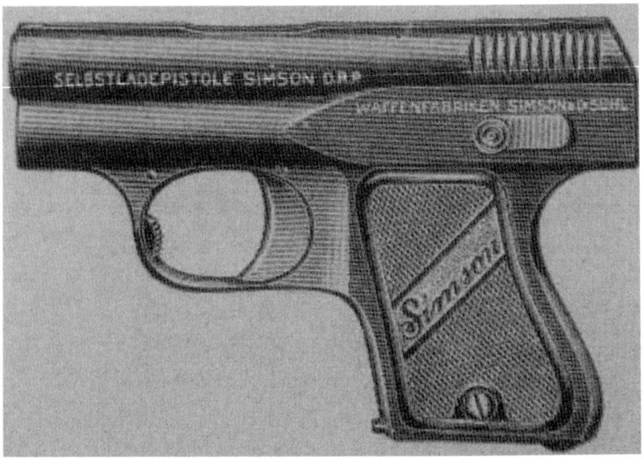

Fig. 155. An early illustration of a pocket pistol marketed by Simson & Co., Suhl. *Courtesy the late William R. Smith.*

Fig. 156: Simson's letterhead showing the company's later trademark and the expansive facilities. *Courtesy Paul Shomper.*

Fig. 156a: A .22 caliber DSM training rifle stamped with the same Simpson trademark. Simson's letterhead showing the company trademark and the expansive facilities. *Courtesy Dr. Geoffrey Sturgess.*

power, were arrested in 1935, and then fled in 1936 at first to Switzerland, and then to the United States. The P08 machines, tooling and fixtures were transferred once again to H. Krieghoff, Suhl. What remained of the company was eventually incorporated into the Gustloff-Werke foundation. Over the years Simson used a number of trademarks, to include a diagonal script "Simson," an upper case "S" with a single triangle, and another using the "S" and a combination of three triangles, first registered in 1918. See also *Gustloff*.

SINOXID: Originally a tradename for rust-free priming composition made by Rheinisch-Westfälische Sprengstoff Aktiengesellschaft. In 1901, RWS started development of rust-free propellants. Primers of the day were traditionally made with potassium chlorate, which often caused rust formation and barrel corrosion. By 1910, increasing numbers of slow-rusting propellants came into use, but the main breakthrough only came in 1918 when the non-corrosive. Trizinate was tested in primers. RWS patented the formula in 1926, and named it SINOXID. DWM also produced it under license, but marketed the composition under the NICORRO tradename. The SINOXID tradename and trademark were used to market ammunition made with RWS' primers in the 1930-1950s. After WWII, the tradename and trademark passed to Dynamit Nobel AG, Troisdorf, and then to RUAG Ammotec GmbH, Fürth. See also *Rheinisch-Westfälische Sprengstoff Aktiengesellschaft* and *DWM*.

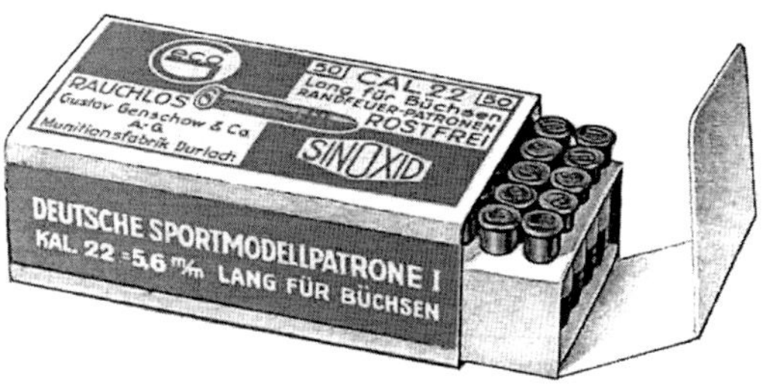

Fig. 157: An advertisement for Geco's "SINOXID" brand of .22 caliber ammunition from ca. 1934.

Sirius: Tradename used by Albrecht Kind, Berlin-Nürnberg to market accessories and rifle telescopes made by other optical firms prior to WWII. See also *Kind, Albrecht.*

Škoda: Škoda Aktiengesellschaft (eh. Skoda-Werke, or Czech *Škodovy závody*), Pilsen[47]; one of the largest industrial conglomerates in Austria-Hungary and later Czechoslovakia. Under German occupation, Škoda was involved primarily in the production of machineguns, mortars, heavy cannon, trucks and other vehicles, but also had a hand in assisting, backing or financing various other Czech small arms enterprises.

SKOPAR: Tradename used on scopes made by Voigtländer & Sohn, Braunschweig prior to WWI. See also *Voigtländer.*

SKOPARETTE: Tradename used on scopes made by Voigtländer & Sohn, Braunschweig prior to WWI. See also *Voigtländer.*

Slavia: Tradename used by the pistol designer Antonin Vilímec, Kdyne, Bohemia. Vilímec essentially eliminated the grip safety of the Browning-designed Colt 1908 and began making and selling the pistols as the "Slavia" (1920-1930s). In 1935 the company was acquired by the Czech company Posumavska Zbrojovka Kohout & Spol. (SPK), Kdyne. SPK continued to manufacture the pistol as the MARS, which was in turn copied and produced under the DUO tradename. The tradename was also used to market air pistols made in Czechoslovakia after WWII. See also *Posumavska Zbrojovka Kohout & Spol. (SPK).*

Sodia: Franz Sodia, Ferlach; custom gunmaker, operated around 1910-1920s.

Soldaten: Tradename used by Gustav Voss, Solingen to market pocket knives 1930-1940s.

Solinger Axt und Hauerfabrik: Solinger Axt und Hauerfabrik, Solingen; maker of cutlery and daggers during the Nazi era.

Solinger Metallwaffenfabrik: Solinger Metallwarenfabrik, Stoecker & Co., Solingen; maker of cutlery, gravity knives and daggers during the National Socialist era. The company was also known as "SOMEFA" and "SMF."

Solothurn: Waffenfabrik Solothurn AG, Solothurn, Switzerland; a front company for Rheinmetall AG to circumvent the restrictions in the Treaty of Versailles and develop and market its products between the wars. Rheinmetall acquired 90% of the Swiss firm in 1929.

SOMET: Tradename of a subsidiary company believed to have been owned by the Czech Meopta, Prerau firm.

SOMEFA: Solinger Metallwaffenfabrik, Stoecker & Co., Solingen; maker of cutlery, gravity knives and daggers during the National Socialist era- also known as "SMF."

SMF: Abbreviation and tradename for the Solinger Metallwaffenfabrik, Stoecker & Co., Solingen; maker of cutlery, gravity knives and daggers during the National Socialist era- also known as "SOMEFA."

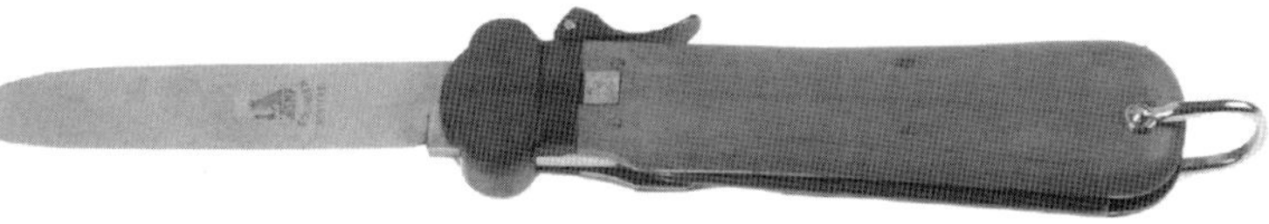

Fig. 158: A Fliegerkappmesser (Flyers Utility Knife) and a close-up of the etched "SMF" trademark owned by the Solinger Metallwaffenfabrik, Stoecker & Co., Solingen. *Courtesy the late William R. Smith.*

Fig. 159: WWI Gewehr 98 as manufactured by the Spandau arsenal. *Courtesy Robert Jensen.*

SPANDAU: Königlich-Preussische Gewehrfabrik Spandau, Berlin-Spandau; the largest and most important Imperial German, state-owned small arms production facility/arsenal prior to the end of WWI. The facility made around a million Gewehr 98s, and thousands of Kar.98s and MG 08/15s. After WWI, the Spandau Arsenal was ordered dismantled due to the Treaty of Versailles. Deutsche Werke, and later the Deutsche-Industrie-Werke were largely established from its remnants. See also *Deutsche-Industrie-Werke*.

Spateneder: Spateneder, München; maker of cutlery and daggers during the Nazi era.

Spindler & Hoyer: Spindler & Hoyer, Göttingen, maker of precision optics, binoculars and rifle telescopes WWI-WWII. The firm was founded in 1898 by August Spindler and Adolf Hoyer. Spindler & Hoyer used a number of tradenames over the years and is best-known for producing thousands of examples of the WWI-era "Fernglas 08."

Fig. 160: A pair of 6 x 30 binoculars made in the 1930s by Spindler & Hoyer, Göttingen.

Spitzer: Gustav Spitzer, Solingen; maker of cutlery and daggers during the Nazi era.

SPK: Abbreviation for Posumavska Zbrojovka Kohout & Spol., Czechoslovakia; firearm manufacturer. See *Posumavska Zbrojovka Kohout & Spol.*

Spreewerk: Spreewerk GmbH, Metallwarenfabrik, Berlin-Spandau; small arms manufacturer originally known as Waffenwerke Oberspree Kornbusch & Co., Berlin-Spandau. During WWI the predecessor of the firm was a DWM subsidiary and made various products to include G98 rifles. In 1935 Deutsche-Industrie-Werke, Berlin acquired a controlling interest in the company, which was subsequently renamed Metallwarenfabrik Spreewerk, Berlin-Spandau. Spreewerk had a proven track record manufacturing cannon and small arms components under license prior to and during WWII, and was additionally tasked to make P38 pistols and MP44 components. When the Volkswaffen Programm was announced in 1944, D-I-W and Spreewerk were well positioned for contracts, as SS-Standartenführer Erich Purucker, charged with arming the Waffen-SS as well as the Volkssturm, just happened to be a former executive in the Spreewerk firm. After several prototypes were tested, D-I-W's example was indeed accepted and contracts soon followed. Receivers were produced by D-I-W in Berlin and Spreewerk manufactured other parts and then assembled the receivers and components into complete rifles at its factory in Grottau, Sudetenland (Czechoslovakia). This was similar to the arrangement Spreewerk developed with its earlier P38 pistol order, and it is thought that the start up of Volksgewehr production essentially halted the production of "cyq" marked P38 pistols the firm had been making in Grottau.[48] Sometime after February 9, 1945, the workers involved and the final assembly were reportedly transferred to yet another facility, the nearby "Georg Elster Werke." Small components were apparently made at Spreewerk, Kratzau and stocks were cut and finished by young apprentices of the firm. See also *Deutsche-Industrie Werke*.[49,50, 51]

Fig. 161: Spreewerk logo as printed on a company identification badge. *Courtesy Jan Balcar and Ron Clarin.*

Springer: Tradename used in conjunction with a pole-vaulter logo- used by Carl Wusthof (Gladiatorwerk) AG, Solingen; maker of cutlery and dress daggers during the National Socialist era.

SRB A ŠTYS: Srb & Štys Fabrik Präziser Messinstrumente, Prag; pre-WWII Czech optical company that made telescopic rifle sights and other optics under German supervision during WWII. Founded in 1919, Srb & Štys marketed some of its rifle telescopes under the "Lovec" tradename. The company was later absorbed into the "people's enterprise" Meopta Košíre, Prerov after WWII.

SS: Not a tradename, but an abbreviation of the "Schutzstaffel der NSDAP." Originally formed as the personal bodyguard of Nazi dictator Adolf Hitler (Stoßtrupp Adolf Hitler), the organization grew during the regime to include all of the German police, security services, and concentration camps, and eventually had its own armed military divisions, or Waffen-SS. The SS "runes" are frequently seen engraved or stamped on small arms and other military items.

S & S: Trademark/abbreviation used by J. P. Sauer & Sohn, Suhl. See *Sauer*.

Staatliche Waffenfabrik, Chatellerault, France; made K98 bayonets under WWII German occupation.

Staehle: C.A. Staehle, Stuttgart; a distributor of cutlery and other items, circa 1930s.

STAR: "STAR" Star Bonifacio Echeverria SA, Eibar, Spain; a Spanish small arms manufacturer. The patriarch of the firm was Jose Cruz Echeverria, who made muzzle-loading arms in the 1800s. His two sons, Julian and Bonifacio Echeverria, took over the company, and by WWI were making "Star" and "Ruby" pistols for the French military. The firm's capacity, however, could not keep up with the demands of the contract, and ultimately a great number of pistols were sub-contracted to Gabilondo y Urresti and some 30 other firms in and around Eibar. The "Star" tradename was first registered in 1919. Between the wars, the company continued to make knock-off versions of the Browning 1903 and 1911 designs and supplied arms to the Republican forces during the Spanish Civil War. When Francisco Franco and the Nationalists wrested power from the Republicans, Star supplied weapons to the new government. During WWII, Star supplied around 30,000 Model B pistols to the German Wehrmacht under contract. It also enjoyed commercial success selling small arms to other Axis and aligned states during the same period as well.

Steigleder: Ernst Steigleder, Suhl & Berlin; maker/retailer of sporting arms in the 1920-1930s.

Steinbach: Julius Steinbach; an edged weapons maker during the National Socialist era.

Steinberg: F. von Brosy Steinberg, Solingen; WWII era cutlery maker.

Steinberger: Julius Steinberger, Ohligs; an edged weapons maker during the National Socialist era.

Steiner: Walther Steiner Eisenkonstruktionen, Suhl; a fairly well known manufacturer of small arms components. Steiner was also apparently tasked to manufacture MP.3008s (simplified Sten design) late in WWII.

Steinheil: Optische-Werke C. A. Steinheil Söhne G.m.b.H., München; maker of optics.

Steinhoff: Franz Steinhoff, Solingen-Wald; maker of cutlery and daggers during the Nazi era. Steinhoff used the tradename "MOBIL" to market some of its products.

Stenda: Stenda Werke Waffenfabrik, Suhl; maker of small 7.65mm pocket pistols such as the "Stenda" (aka "Beholla") from about the early to mid-1920s.

Stereor: Brand name used on some optics manufactured and marketed by Hensoldt prior to WWII under the "Teljagd" series.

Stern: Tradename of used to market a conventional 6.35mm pistol made by Albin Wahl, Zella-Mehlis in the 1920-1930s. Stern pistols were stocked in grips bearing the Star of David trademark and the company abbreviation "A.W.Z."

Steyr: Steyr-Daimler-Puch AG, Steyr Austria; a large conglomerate and manufacturer of rifles, pistols and other military items. The company was founded in 1863 and was renamed Österreichische Waffenfabriks-Gesellschaft STEYR in 1869. Steyr had some 15,000 employees and produced a number of Mannlicher-designed rifles, pistols and other military goods prior to and during WWI. Facing bankruptcy after the war, ÖWG branched out into automobiles and eventually was merged with two other auto makers, Austro-Daimler and Puch. During this same time, the German Rheinmetall firm was in large part controlled by the "Hermann Göring-Werke" foundation. By the mid-late 1930s, Rheinmetall had in turn acquired a large percentage of Steyr stock. In regard to small arms between the world wars, Steyr then entered into agreement to manufacture Rheinmetall-designed weapons, which were in turn marketed abroad via the Swiss front company Solothurn. When the Nazis annexed Austria in 1938, Steyr became more overtly and tightly controlled by Hermann Göring-Werke foundation. Steyr produced hundreds of thousands of small arms (in addition to automobiles, trucks and other goods) for the Wehrmacht during WWII, most notably sub machineguns, MP44s, K98s and VK98s. Though based and headquartered in Steyr, the company had subsidiaries during the war in Graz, Radom, Warschau (Warsaw), Wien, St. Valentin and other locales.[52] Steyr-Daimler-Puch AG was dismantled after WWII but was recreated later, and is still a major manufacturer of sporting and military weapons at the time of this writing. See also *Österreichische Waffenfabriks-Gesellschaft STEYR*.

Fig. 162: Steyr's original method of marking its small arms as shown on this M.1895 Rifle. *Courtesy Robert Jensen.*

Fig. 163: The cover of Steyr's 1935 commercial firearms catalog.

Fig. 164: A simple newspaper advertisement from the 1940s making use of the firm's circular "target" trademark.

Fig. 165: Details from an MP34(ö) made by Steyr. The weapon bears the firm's traditional trademark as well as another target-type trademark that is currently thought to represent a facility in Poland, supervised by Steyr. Another similar logo has been observed with an additional "W" in the center and is also thought to represent "Steyr-Daimler-Puch, Warschau." *Courtesy Dr. Geoffrey Sturgess.*

Fig. 166: In the middle of WWII, Steyr built a prototype rifle intended to replace the K98. This rarely seen "G.42(St)" bears the company trademark well after arms makers were ordered to stop the practice; as it was only meant to be presented to German military testing and acceptance officials.

StG: Not a trademark, but an abbreviation seen on barrel blanks supplied by Steyr-Daimler-Puch AG. See also *Steyr*.

Stiegele: Carl Stiegele Jr., München; a distributor and retailer of small arms.

Stock: G. u. W. Stock, Eisenach; an edged weapons maker during the National Socialist era.

Stock: Franz Stock, Maschinen- u. Werkzeugfabrik, Berlin-Neukölln; maker of various pistols prior to and circa WWII.

Stock: Gustav Stock; an edged weapons maker during the National Socialist era.

STOCKO: "Stocko" Metallwarenfabriken OHG, Wuppertal-Elberfeld; primary maker of non-disposable belts for MG42 ammunition, but also thought to have made a small number of VG1 rifles for the Volkssturm near the end of WWII.

Stoewer: Stoewer-Werke AG, Kraftfahrzeugfabrik, Stettin; founded in 1858 by Bernhard Stoewer as a sewing machine company that later transitioned into bicycle and automobile manufacturing. Stoewer is thought to have made a small number of VG1 rifles for the Volkssturm near the end of WWII.

Stöver: Otto Stöver; Solingen; WWII era cutlery maker.

Stübgen: Fr. Stübgen & Co., Metallwerk, Erfurt; a manufacturer of metal goods, known to have made various machinegun mounts, cases and Flak magazines during WWII.

Fig. 167: Detail from a WWII-era Flak 30 magazine made by Fr. Stübgen & Co., Metallwerk, Erfurt and marked with the company's peculiar "bat" trademark. *Courtesy Dr. Geoffrey Sturgess.*

Stukenbrok: August Stukenbrok, Einbeck; large wholesaler and catalog dealer of mostly German and Belgian ammunition, firearms and accessories, circa 1900-1915 and beyond.

ST. W.: Abbreviation for Stenda Werke Waffenfabrik, Suhl; maker of small 7.65mm pocket pistols, such as the "Beholla" from about the early to mid-1920s.

Südbahn-Werke: Südbahn-Werke, AG, Wien; made a small number of VG1 Volksgewehr at the end of WWII.

Süd: Süd Messerfabrik, Sud; an edged weapons maker during the National Socialist era.

Süddeutsche Messerfabrik: Süddeutsche Messerfabrik, Gefrees; WWII era cutlery and dagger manufacturer.

Südmetall: Südmetall AG, Mußbach/Weinstraße; made stamped metal components as well as G/K43 rifle and other magazines during WWII.

Suhl: Waffenstadt Suhl, Suhl; an organization made up of several manufacturers of training rifles in and around Suhl. Some companies supplied components, or stocks, and others simply assembled components into complete rifles. J.P.Sauer, C.G. Haenel and BSW were prominent participants, as was Krupp, Essen (steel for barrels).

Suhl Consortium: A group of small arms makers in Suhl that collaborated to make rifles and pistols during the Imperial era. The major participants were C.G. Haenel, J.P. Sauer & Sohn, and V. Chr. Schilling.

Suhler Waffengesellschaft: Suhler Waffengesellschaft, Suhl; small arms manufacturer in the early 1900s.

SuS: Abbreviation for J.P. Sauer & Sohn Gewehrfabrik, Suhl. See *Sauer*.

Süss: N. Süss, Budapest; manufactured rifle telescopes for the Austro-Hungarian Army circa WWI.

Swarovski: Daniel Swarovski, D., Glasfabrik u. Tyrolit-Schleifmittelwerke, Tyrol, Austria; maker of precision optics and rifle scopes during WWII. Founded by Daniel Swarovski (1862-1956) in 1895, the company became famous for fine crystal and cutting tools. Wilhelm Swarovski established the optical subsidiary in 1935. The company marketed some of its products under the "Habicht" tradename. Swarovski Optik was still in business at the time of this writing.

Fig. 168: A .22 caliber training rifle made up from components supplied by several manufacturers in and around the city of Suhl. Well over 10,000 .22 caliber training rifles were made under the "Waffenstadt Suhl" umbrella between about 1934-1937. *Courtesy Robert Jensen.*

Tannenwerk: Tannenwerk GmbH; a maker of edged weapons during the Nazi era.

Televist: Tradename used by Carl Ruhnke, Berlin to market optics. See *Ruhnke*.

TELL: Tradename used by Cuno Büchel, Suhl to market sporting arms between about the 1880-1920s era. Adolf Frank, Hamburg is also known to have marketed/distributed items, such as telescopes, under this brand as well. Tradename was acquired by AKAH after WWII. See also *AKAH*.

TESCO: Tradename/company name derived from the names "Teschner" and "Collath." See *Collath*.

Thälmann: VEB Fahrzeug- und Jagdwaffenwerk Ernst Thälmann, Suhl: A post-WWII German Democratic Republic (GDR/DDR) state-owned consortium made up from the remnants of a number of wartime small arms makers, including Haenel, Gustloff, Krieghoff, Sauer, and the Fortuna-Werke. The concern was named after the head of the German Communist Party (KPD) who was arrested by the Nazis and eventually executed near the end of WWII. VEB Thälmann made sporting and military products and small arms for the GDR/DDR, small arms and other military products for export abroad.

33 841552 V 10973

Fig. 169: The VEB Fahrzeug- und Jagdwaffenwerk Ernst Thälmann, Suhl trademark as it appears in the records of the *Deutsches Markenamt*.

THERAR: Tradename used by Emil Busch AG, Rathenow, used to market spotting scopes prior to WWII. See also *Busch*.

Thieme & Schlegelmilch: Thieme & Schlegelmilch; manufacturer of Drillings and sporting weapons.

THÜRINGEN: Waffengenossenschaft Thüringen; an organization made up of several manufacturers of training rifles in the Thüringen region. Some companies supplied components, or stocks, and others simply assembled components into complete rifles in the 1930s.

Fig. 170: The heavily-forested arms making region of Thüringen billed itself as the "Green Heart of Germany" and used this logo in tourist brochures to promote the region.

Tiegel: Karl Tiegel (Tiegelwerk), Riemberg-Breslau; WWII era cutlery maker.

Tiegelwerk: Karl Tiegel, Riemberg-Breslau; an edged weapons maker during the National Socialist era.

TIGER: Tradename used on knives made by Tigerwerk, Lauterjung & Sohn GmbH & Co., Solingen. This same company used the "Puma" tradename on other lines of cutlery.

Tillmanns: Carl Tillmanns & Söhne, Solingen; maker of cutlery and daggers marketed under the "LUX" tradename during the Nazi era.

Timner: AFW Timner, Coblenz; manufacturer of Drillings and other sporting weapons.

TIP-TOP: Tradename used to market rifle telescopes made by Karl Kahles, Optiker, Wein.

Fig. 171: A group of firearms manufacturers used a variation on this same regional logo to market small caliber training rifles. *Courtesy Robert Jensen.*

Tiroler Waffenfabrik: Tiroler Waffenfabrik Johann Peterlongo, Innsbruck, Austria; large manufacturer and catalog retailer of firearms and accessories founded in 1854. The company was acquired by Richard Mahrholdt & Sohn in 1939 and remained in business well after WWII.

Torley: Gebr. Torley, Solingen; a WWII-era ceremonial dagger manufacturer.

Torpedo: Tradename used by the Torpedo-Werke, Frankfurt as well as Fichtel & Sachs, Schweinfurt. See also *Fichtel & Sachs* as well as *Torpedo-Werke*.

Torpedo-Werke: Torpedo-Werke AG, Fahrräder u. Schreibmaschinenfabrik, Frankfurt/ Main-Rödelheim; a manufacturer of bicycles and typewriters that also was pressed into making small arms components during WWII.

Troost: Troost-Wolf; an edged weapons maker during the National Socialist era.

Tückmantel: Gebr. Tückmantel; an edged weapons maker during the National Socialist era.

Uhlenflucht: Tradename probably used before and after WWII by Robert Geller, Giessen; a maker of telescopic rifle sights from ca. the 1930s.

Ullrich: Carl Ullrich & Co. Metallwaren, Oberschönau, bei Zella Mehlis; a traditional maker of hand tools founded in 1883. The firm was a principal supplier of K98 rifle components and other military goods during WWII. Ullrich marketed tools and other goods under the "CULCO" tradename.

Undine: Undine, Solingen; manufacturer of dress daggers during the Nazi era.

Unger: Joseph Unger, Graz; made high quality shotguns and sporting arms circa 1920-1930s.

UNIQUE: Tradename used to market Model R17 "Ruby" type 7.65 pistols made by Manufacture d'Armes des Pyrenees Francaises (MAP), located in Hendaye, France. MAP began making the pistols in the 1920s, and after the fall of France in 1940 continued making them under German occupation until overrun by the Allies. The Germans subsequently redesigned the Unique by altering the rear grip and adding an external hammer. These pistols became known as the Kriegsmodell Rr51.

URANOS: Tradename used by Schütz, Ruf & Co, Kassel, to market 10x binoculars prior to the Second World War.

V: See *Vollmer*.

VAF/C: See *Vereinigten Automaten Fabriken*.

V.C.S.: Abbreviation used by Valentin Christoph Schilling, Suhl. See *Schilling*.

VDE: Not a trademark or tradename at all, but a stylized representation of the *Verband Deutscher Elektrotechniker* (German Federation of Electrical Engineers)-this mark frequently seen on items made from thermoset plastics, or "Kunststoffe."

VDM: An abbreviation for Vereinigte Deutsche Metallwerke. See *Vereinigte Deutsche Metallwerke*.

V.E.B.: Not a tradename, but rather an abbreviation for Volkseigener Betrieb, or "People's Enterprise." Many WWII wartime companies were reorganized or consolidated into V.E.B. after the war in communist East Germany (German Democratic Republic).

VEB Augenoptik "Hermann Duncker": Augenoptik "Hermann Duncker," Rathenow: See *Busch, Emil*.

VEB Fahrzeug- und Jagdwaffenwerk Ernst Thälmann: Fahrzeug- und Jagdwaffenwerk Ernst Thälmann, Suhl: See *Thälmann*.

VEB Rathenower Optische Werke: VEB Rathenower Optische Werke, Rathenow. See *Emil Busch*.

VEB Carl Zeiss: VEB Carl Zeiss, Jena: See *Zeiss*.

Fig. 172: The Venditor tradename molded into a phenolic binocular case made by Venditor Kunststoff-Verkauf GmbH, Troisdorf. The designation on the left is a code for the type of pressing composition, with the numeral "43" representing the company. A thorough discussion of theses codes can be found in *Kunststoffe: A Collector's Guide to German World War II Plastics and their Markings*, available from Schiffer Military Books.

Venditor: Tradename of the Dynamit A.G. subsidiary, Venditor Kunststoff-Verkauf GmbH, Troisdorf; maker of binocular cases, flare cases, other military components.

Venus: Oskar Will, Venus-Waffenwerk; Zella-Mehlis, Thüringen; manufacturer of sporting arms, components, barrels, air rifles and small caliber training rifles. Oskar Will sold out to Wilhelm Foss, Zella-Mehlis in 1921, and the consolidated company was later renamed "Venus Waffenwerk." Products made by Venus were also marketed under the "MARS" tradename. During the latter half of WWII, Venus made MP44 and other small arms components.

Vereinigten Automaten Fabriken: "VAF" Vereinigten Automaten Fabriken Pelzer und Cie, Cöln-Ehrenfeld; a manufacturer of vending machines, pressed into service during WWI to make P08 snail drum magazines (though the company is not thought to have completed many, if at all).

Vereinigte Deutsche Metallwerke: Vereinigte Deutsche Metallwerke AG, Zweigniederlassung der Süddeutschen Metallindustrie, Werk Nürnberg; a large metal processing conglomerate created in 1930 by the merger of the Heddernheimer Kupferwerk and the Süddeutsche Kabelwerk AG, under the management of the Metallgesellschaft AG. "VDM" employed over 21,000 workers by 1934. During WWII the firm was heavily involved in the production of propellers for the Luftwaffe, but in the context of this reference, was a manufacturer of MG17 triggers and other small arms components.

Vereinigte Oberschlesische Hüttenwerke: Vereinigte Oberschlesische Hüttenwerke AG, Werk Andreashütte, Andreashütte; a heavy-industry steel manufacturer that also had some involvement in Volksgewehr production near the end of WWII.

VIS: Abbreviation/semi-tradename used by Fabryka Broni W. Radomiu, Radom, Poland. See *Radom*.

VISAR: Tradename used by Emil Busch AG, Rathenow in marketing telescopic rifle sights prior to WWII.

Vitting: Eduard Vitting, Solingen; WWII era cutlery and dagger manufacturer.

Voigtländer: Voigtländer & Sohn AG, Optisches Gerät, Braunschweig; a longtime leader in the development and manufacturer of precision optics. The company was actually founded in Wien (Vienna), Austria, in 1756 by Johann Christoph Voigtländer. Initially the company was primarily involved in producing scientific instruments, but under the direction of the founder's son, started producing optical lenses and opera glasses. Voigtländer's grandson, Wilhelm Friedrich, took the lead in the company's production of cameras in 1840, and then nine years later the entire operation was moved to Braunschweig, Germany. The firm made a wide range of optical equipment during both world wars, to include binoculars, gun cameras, and telescopic rifle sights. Voigtländer used a number of trademarks and tradenames over the years, including "Scoparette," "Scoparillo" and others. During WWII, the head of the firm was Dr. Ing. Max Ruhland and its chief designer was Dr. Baumgärtner. Baumgärtner had a profound influence on the design and development of the ZF4 rifle telescope as used on the G/K43 series of rifles and experimentally on K98s and MP44s. The company survived the war, but Zeiss eventually acquired a majority of the firm's shares and absorbed it in the 1960s.

Fig. 173: A mid-late 1930s variant of the trademark used by Voigtländer & Sohn AG, Optisches Gerät, Braunschweig, as engraved on a set of 6x30 binoculars.

Völker: Adolf Völker, Zangenfabrik, Schmalkalden; WWII era cutlery and dagger manufacturer.

Vollmer: Not a tradename, but rather refers to the famous small arms designer, Heinrich Vollmer (1885-1961). Heinrich Vollmer established his own company in his mid-twenties, the Vollmer Werke GmbH, Biberach an der Riß in 1909. Vollmer worked with Mauser and Geco between the wars in several attempts at providing the Wehrmacht with machineguns and an early intermediate cartridge assault rifle. Later, Vollmer became the chief designer at ERMA, Erfurt and developed several innovative weapons over the course of his career. These included the VPM 1930, the 7.5mm M35 Automatic Karabiner, the EMP, and the outstanding MP38 and improved MP40 sub machineguns. Vollmer's original company in Biberach was still in the business of manufacturing cutting tools and saw blades at the time of this writing.

Vollmer Werke: Vollmer Werke GmbH, Biberach an der Riß; see *Vollmer*.

VOM HOFE: Tradename used by W. Gehmann of Karlsruhe & Stuttgart, named for the originator of Vom Hofe high velocity cartridges, about 1937.

Von Brosy: F. von Brosy Steinberg, Solingen; cutlery and dagger manufacturer during the WWII era.

Von Lengerke & Detmold: Von Lengerke & Detmold, New York; a major distributor and retailer of small arms. This firm had exclusive rights to import Mauser firearms into the United States for many years.

Voos: Emil Voos Waffenfabrik, Solingen; cutlery and dagger manufacturer during the National Socialist era.

Voss: Gustav Voss, Solingen; cutlery and dagger manufacturer during the WWII era. Voss used various brand and tradenames over the years to include "Soldaten."

Waffengenossenschaft Thüringen: Waffengenossenschaft Thüringen; a conglomeration of small arms makers in the Thüringen region, best known for making small arms components and small caliber training rifles during the National Socialist era. See also *Thüringen*.

Waffenhammer: Max Kaiser Waffenhammer, Deggendorf; cutlery and dagger manufacturer during the National Socialist era.

Waffen-Jung: Waffen Jung, Stuttgart; made .22 caliber training rifles based on the Model 1871 action. Also distributed rifles made by other manufacturers.

Waffenstadt: Waffenstadt Suhl, Suhl; an organization made up of several manufacturers of training rifles in and around Suhl. See *Suhl*.

Waffenwerke Brünn: WWII German name for Ceskoslovenska Zbrojovka, Brno; aka Waffenwerke Brünn I, Werk Brünn. See *Ceskoslovenska Zbrojovka*.

Waffenwerke Brünn: WWII German name for Považská Bystrica, Bystrica; aka Waffenwerke Brünn II, Werk Bystrica. See also *Považská Bystrica* and *Ceskoslovenska Zbrojovka*.

Wagner: I.C. Wagner & Co. GmbH, Heyerode; manufacturer of various metal goods to include StG 44, FG 42, MG 81, and MG 131components during WWII.

Wagner: Wagner & Lang, Solingen; an edged weapons maker during the National Socialist era.

Wagner: Wilhelm Wagner, Solingen-Merscheid; cutlery and dagger manufacturer during the National Socialist era.

Wagner & Lang: Wagner & Lang, Solingen; cutlery maker during National Socialist era.

Wahl: Albin Wahl, Zella-Mehlis; made and marketed conventional 6.35mm pistols using the "Stern" brand name, and trademarked Star of David and "A.W.Z." grip plates.

WALKURE: Tradename used to market small pistols with conical barrels made between ca. 1900-1930 by Oskar Will, Venus-Waffenwerk; Zella-Mehlis.

Wallmeyer: August Wallmeyer, Eisenach-Thüringen; made K98 bayonet scabbards.

Walter: Otto Walter; a maker of 8x40 binoculars in the Imperial era or perhaps post-WWI. Little is known about this firm except that it was likely located in Fürstenfeldbruck. Further research is clearly warranted.

Walter: Thüringische Maschinen- und Fahrradfabrik Walter & Co. GmbH, Mühlhausen i/Thür.; a maker of bicycles, motorcycles and bayonets during the Imperial era and beyond. The firm was also known as the "Möve-Werke, Walter & Co. GmbH, Mühlhausen i/Thür."[53]

Walther: Carl Walther Waffenfabrik, Zella-Mehlis, Thüringen. The Walther firm was founded in 1886 by Carl Walther in Zella-St. Blasii, Thüringen (the town's name was later changed in 1919 to Zella-Mehlis). Walther's first products were target rifles based on the Martini system. Carl Walther fathered five sons during this time. Fritz, Georg, and Willy all followed their father's lead and entered the gun-making business, while Hans became a businessman and Carl Lothar a tool maker. By the turn of the century, Walther had expanded the company and its products. Fritz was apprenticed at the Ludwig Loewe firm in Berlin for a while, and as a result of his experiences and experiments there convinced his father to start producing pistols in Zella-Mehlis. The Model 1 pistol commenced production as a result in 1908. In 1912, the company registered its first version of its famous "Walther banner" trademark.[54] Carl Walther died in 1915, but his initials adorned another of the firm's registered trademarks, the early "CW in an oval" approved in 1917. After his father's death, Fritz was made head of the company, Georg served as the General Director, and Hans was appointed to head the sales department. Facilities were greatly expanded during WWI, with Walther supplying pistols, MG08 components and other products to the German armed forces. During the 1920s, Walther stayed afloat by making bicycle parts, adding machines, flare guns and sporting weapons. Then in 1929 Fritz Walther began to market his now-famous "PP" and "PPK" pistols, both of which became popular with civilians, the police and the National Socialists. Indeed, Walther flourished under the Nazi regime and received large government and para-military contracts to produce training rifles, PP/PPKs, P38s, G/K43 rifles and Volksgewehre. Some of these products or components were made at Metallwerke Neuengamme GmbH, Hamburg-Neuengamme, which was a subsidiary within the Neuengamme KZ. By the end of the war, Walther employed 2,500 workers in Zella-Mehlis alone. Unfortunately, Zella-Mehlis fell inside the Soviet zone of occupation and Fritz Walther was forced to flee westward. The Zella-Mehlis facilities were demolished, and most of the records and tooling disappeared behind the Iron Curtain. After a short stay in U.S. military custody, Fritz reestablished the firm in the American zone of occupation, at first returning to the manufacture of adding machines. He then licensed some of his pistol designs to other firms, such as Hämmerli AG in Switzerland and Manurhin in France. Subsequently, Walther began making air rifles and pistols again and acquired facilities in the city of Ulm. Waffenfabrik Walther used a number of trademarks over the years-from a stylized "CW," "CeWe," and its famous "banner." Fritz Walther moved quickly to reregister these trademarks and even claimed ownership of the streamer itself. By 1953 the firm was running strong, making PP & PPK pistol components for Manurhin, blank firing pistols, and other products. In 1957, Walther received a contract from the fledgling German Bundeswehr to produce the P1 pistol, an updated version of the P38. Fritz Walther died in 1966, but the company remained in business and under family control for many years. In 1993, Carl Walther GmbH was acquired by the Umarex Group and remains in business at the time of this writing.

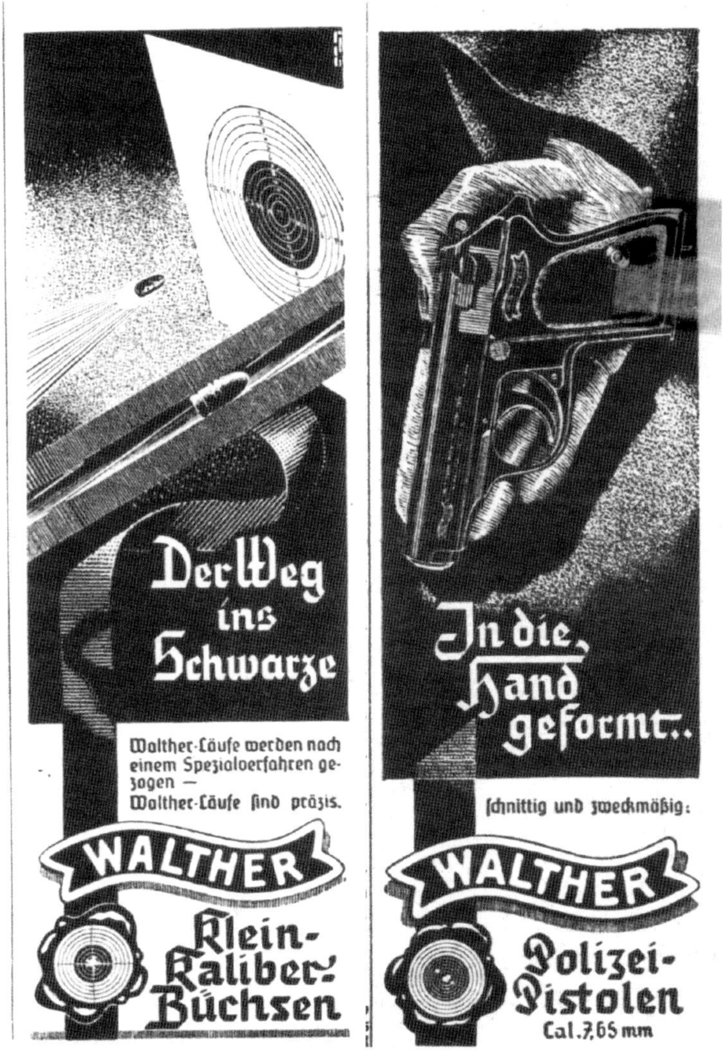

Fig. 174: Advertisements from the 1940s showing the familiar Walther "banner" and hugely successful KKW training rifle and its PPK pistol.

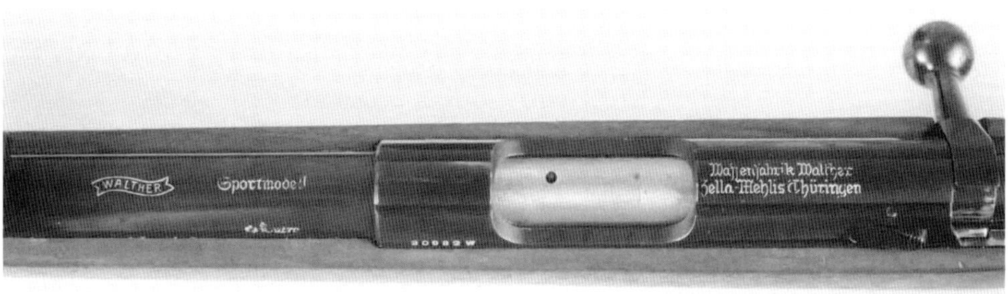

Fig. 175: A nice example of the Walther .22 caliber "Sportmodell" stamped with the company trademark, name and location.

Fig. 176: A nice example of a late WWII Walther PP with the firm's trademark pressed into hardwood grip panels.

107918

CARL WALTHER WAFFENFABRIK

AUTOMATISCHE WAFFEN · RECHENMASCHINEN

WALTHER

Fernschreiber 068 184

Fernsprecher: Nr. 651 und 652 Waltherwerk
Telegramm-Adresse: Zellamehlis
Codes: Rudolf-Mosse-Code; ABC 5. Edition
Liebers 5 Letters; Bentley
Postscheck-Konto: Erfurt Nr. 6520
Bank-Konten: Reichsbank-Girokonto, Suhl
Thüringische Staatsbank, Zella-Mehlis
Stadtsparkasse Zella-Mehlis
Girokonto Nr. 211
(15)
ZELLA-MEHLIS (THÜR.)
Postschließfach 41

An den
Reichsführer SS
Befehlshaber des Ersatzheeres
Stabsführer des Deutschen
Volkssturmes, Beauftr.f.Bew. u. Ausrüstung
SS-Standartenführer P u r u c k e r
(1) Berlin - Grunewald
Douglasstrasse 7

Fig. 177: Walther's late-WWII letterhead. *NARA*.

Walther: Lothar Walther, Zella-Mehlis; a manufacturer of machine tools and small arms components during WWII.[55] After the war, the firm was reestablished in West Germany and remains in business at the time of this writing, manufacturing barrels and sub-caliber devices.

WASSO: Tradename used by Abraham Schnittert Abr, WASSO, Solingen-Wald to market cutlery and daggers during the Nazi era.

Weber: Fritz Weber & Co., Laternenfabrik, Berlin; traditionally a manufacturer of lanterns. Weber made various ammunition cans and MG accessories during WWII.

Weck: Weck & Stamm, Weyer; an edged weapons maker during the National Socialist era.

Weihrauch: Hermann Weihrauch Gewehr und Fahrradteile Fabrik, Zella-Mehlis; a manufacturer of scope mounts, small arms components and small caliber training rifles. Founded in 1899, Hermann's three sons, Otto, Werner, and Hermann, Jr. joined the company later. Between the wars, the firm made shotguns, .22 caliber training rifles, accessories and bicycle parts. Weihrauch was well known for scope mounts during the Nazi era. After the end of WWII the firm was reestablished in West Germany as Hermann Weihrauch Sportwaffenfabrik, and is known to have used the tradename "ARMINIUS," among others.

Fig. 178: A very rare DSM .22 caliber training rifle made by Herman Weihrauch Gewehr und Fahrrad Teilefabrik, Zella-Mehlis. Weihrauch used a number of trademark variations over the years. *Courtesy Bob Marquard and Paul Shomper*

Fig. 179: Detail of a Hensoldt SS-Dienstglas 4x scope on a first-pattern short side-rail mount also made and marked by Hermann Weihrauch. Weihrauch in fact was the lead design firm for telescopic mounts during the National Socialist regime. *Courtesy Dr. Geoffrey Sturgess.*

Weinrank: Franz Weinrank; manufacturer of edged weapons during the National Socialist period.

Weisgerber: W. Weisgerber, St. Wendel; manufacturer of Drillings and sporting weapons.

Weithmann: Josef Weithmann, Augsburg: a maker of cutlery and dress bayonets during the Nazi-era.

Weltersbach: Wilhelm Weltersbach, Solingen; maker of cutlery and ceremonial daggers during the Nazi-era.

Wiener Waffenfabrik: Wiener Waffenfabrik, Wien, Austria; small arms manufacturer. The company was founded in 1909 by Alois Tomiška and Camillo Frank in order to produce Tomiška's patented pocket pistol, the "Little Tom" (Tomiška actually means "Little Tom" in Czech). Tomiška sold his name and patents to the firm in 1919 and retired to Pilsen. WWF continued production of the pistol into the mid-1920s.

Weiputz: Thomas Weiputz, Solingen-Höhscheid; founded in 1880, the firm was a well known cutlery manufacturer for many decades.

Werner: Fritz Werner AG, Maschinen- u. Werkzeugfabrik, Berlin; a machine, tool and barrel maker, heavily involved in supplying cartridge-making machinery, headspace gauges, milling machines, barrel blanks, and finished barrels to small arms makers during the pre-WWI and WWII eras. The company was founded in 1896 by Friedrich Karl Werner and went public in 1915. After WWI, the firm had significant business relations with Solothurn in Switzerland, and by 1936 employed some 3,000 workers. Werner had a number of factories and offices in and around Berlin and its suburbs. The company was folded into the DIAG concern after WWII and eventually acquired by the M.A.N conglomerate. See also *M.A.N.*

Werth: A. Werth, Solingen. Founded in 1828, Werth was maker of cutlery and ceremonial daggers during the Nazi-era.

Wessel & Müller: Wessel & Müller Beschlagfabrik, located in Luckenwalde, maker of stamped metal optics cases and other sheet metal goods.

Wester & Butz: Wester & Butz, Solingen; a maker of cutlery and ceremonial daggers during the Nazi-era.

Westfalia Werkzeuge: Westfalia Werkzeuge, Hagen, maker of punches, chisels and other small hand tools for the small arms and other industries.

Wetzel: Georg Wetzel; an edged weapons maker during the National Socialist era.

Weyersberg: Gebrüder Weyersberg, Solingen; cutlery maker during the National Socialist era. The firm was in business at least as early as the 1890s and used a number of trademarks and logos over the years based on a trumpet as the main figurative element.

Weyersburg: Gottfried Weyersberg, Solingen; maker of fine cutlery, daggers, knives during the Nazi era.

Weyersberg: Gustav Weyersberg Nachfolger, Solingen; cutlery maker during the National Socialist era.

Weyersburg: Weyersberg, Kirschbaum und Cie., Waffenfabrik GmbH, Solingen-Wald. The Weyersburg and Kirschbaum families have a long heritage in and around Solingen, dating back many centuries. Indeed, a family ancestor, Wilhelm Weyersburg was elected as the mayor of Solingen proper in 1573. Around the same time, the area in and around Solingen became well known for sword production, a business in which both the Weyersburg and Kirschbaum families were heavily involved. Kirschbaum first registered its trademark in 1774 and Weyersburg its famous "Knight's Head" in 1854. In 1883, the two companies merged, forming the most influential business at the time in the city, "WKC." By 1892 the firm was employing over 1,200 workers, and a few decades later was making edgeware for the Kaiser's military. In 1922, WKC was acquired by the "Siegen Solinger Gu□stahlverein," but suffered financially during the Great Depression. The company prospered with the rise of the National Socialists and became a major supplier of swords, daggers, bayonets and other edgeware to the various NSDAP organizations, other government departments and the military. The firm also made bicycles under the "Patria" tradename. Most of WKC's facilities were destroyed by Allied bombing near the end of WWII. After the war, the remaining assets, trademarks and tradename were acquired by Hans Kolping, Solingen and remain in use today.[56] See also *Weyersburg* and *WKC*.

Fig. 180: The title page of WKC's cutlery catalog dating from the 1930s displaying the company's famous "knight head" trademark.

Weyersberg: Waffenfabrik Max Weyersberg, Solingen; another Weyersberg family member involved in making daggers for the National Socialist regime. Max Weyersberg used the "WMW Waffen" trademark and tradename.

Weyersburg: Paul Weyersburg & Co. Waffenfabrik, Solingen; maker of cutlery, daggers, and K98 bayonets during the National Socialist era.

Weyersberg: Reinhard Weyersberg, Solingen; another member of the Weyersberg family involved in dagger manufacturing during the Nazi era.

Wichmann: Wichmann, Gebr., Zeichengeräte, Vermessungsinstrumente, Berlin; an instrument manufacturing company founded in 1873, noted in the context of this reference for making rifle training devices and other military goods during the Nazi era.

Will: Oskar Will, Venus-Waffenwerk; Zella-Mehlis, Thüringen, manufacturer of sporting arms, components and small caliber training rifles. Originally an independent firm, Oskar Will was acquired and incorporated into Venus-Waffenfabrik prior to WWII.

Willers: Fa. Willers, Jena; a small manufacturer of rifle telescopes ca. late 1880s-1910. Further research is clearly warranted.

Wilke: H. Wilke & Co., Remscheid; an edged weapons maker during the National Socialist era.

Wimmersperg: Heinrich Wimmersperg, Berlin-Halensee; designed a double-barrel machinegun, self-loading rifles, and an unusual 7.92mm Kurz bullpup rifle in late WWII which utilized the ZF4 scope. The weapon was not accepted for mass production.

Wingen: Anton Wingen Jr., Solingen; maker of fine cutlery, daggers, knives during the Nazi era. Some of the Wingen products were marketed under the "CHROMOLIT" and "OTHELLO" tradenames.

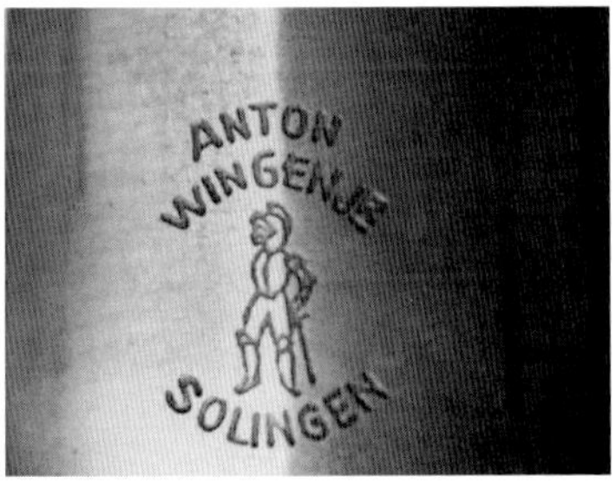

Fig. 181: Anton Wingen's distinctive logo, as etched on National Socialist era cutlery. *Courtesy Robert Jensen.*

Winkel: R. Winkel GmbH Optische Industrie, Göttingen; a maker of precision optics. Winkel was part of the Zeiss conglomerate and used the "WINKEL-ZEISS" tradename.

Wirth: Gustav Wirth, Gräfrath; cutlery and dagger maker during the Nazi era.

Wissner: Metallwarenfabrik AG (eh. H. Wissner), Zella-Mehlis; founded in 1892, made small arms components during WWII. The firm was reestablished after WWII in Bensheim.

Witte: Ernst Erich Witte, Solingen; maker of cutlery and daggers marketed under the "Kroneck" tradename during the National Socialist era.

Witte: P. D. Witte, Solingen; maker of fine cutlery, daggers, knives during the Nazi era.

WKC: Abbreviation/trademark of Weyersberg, Kirschbaum und Cie., Waffenfabrik GmbH, Solingen-Wald, maker of cutlery, daggers, K98 bayonets during 1930-1940s. See also *Weyersberg, Kirschbaum und Cie.*

WMW: Waffenfabrik Max Weyersberg, Solingen; cutlery and dagger manufacturer during the National Socialist era.

Wöhler: Dr. F. A. Wöhler, Kassel, maker of precision optics and rifle telescopes. This firm is known to have used the "DECAMAR" tradename.

Wolf: Fritz Wolf, Rob. Sohn, Gewehrschaftfabrik, Zella-Mehlis, a rifle stock maker.

Wolf: Joseph Wolf Waffenfabrik, München; cutlery and dagger maker during the Nazi era.

Wolf: Karl Wolf; an edged weapons maker during the National Socialist era.

Wolf: Wilhelm Wolf & Sohn; an edged weapons maker during the National Socialist era.

Wolfertz: Isidor Wolfertz, Solingen; cutlery and dagger maker during the Nazi era.

Wolt: Wolt, A. W. Wolt, Suhl, manufacturer of fine sporting arms, around 1930.

WURAG: "WURAG" Eisen- u. Stahlwerke AG, Hohenlimburg; an arms maker during WWII. The company fell under the control of the Krupp concern after WWII.

Würzer: Rudolf Würzer, Christofen; an edged weapons maker during the National Socialist era.

Wüthrich: W. Wüthrich, Werkzeugbau, Lutzelfluh, Schweiz, manufacturer of sporting rifles.

Wüsthof: Gladiatorwerk Carl Wüsthof (Gladiatorwerk) AG, Solingen; maker of cutlery and dress daggers during the National Socialist era.

Wüsthof: Eduardt Wüsthof Dreizackwerk, Solingen; cutlery and dagger manufacturer during the National Socialist era.

WWF: Wiener Waffenfabrik, Wien, Austria; maker of the "Little Tom" pocket pistols. See also *Wiener Waffenfabrik*.

WwSu: Abbreviation for Waffenwerkstadt Spandau; known to have made various rifle headspace and manufacturing gauges during the Imperial era.

Yale & Towne: The Yale & Towne Mfg. & Co., Velbert; made stamped metal components and G/K43 rifle and other magazines during WWII.

Z (within rifled bore trademark): Trademark used by Praga Zbrojovka, Prague and later Ceskoslovenska Zbrojovka, Brno. See *Praga Zbrojovka*.

Z (within a circle trademark): See *ZB, Považská Bystrica.*

Zander: Carl Zander, Solingen; cutlery and dagger maker during the Nazi era.

ZB: ZB, Považská Bystrica**,** Czechoslovakia; made vz.24 rifles for the Czech military and for foreign sales before WWII. When Czechoslovakia was annexed by the National Socialists, the facility was renamed Waffenwerke Brünn II, Werk Bystrica, and spent the rest of the war making K98s for the German Wehrmacht. See also *Ceskoslovenska Zbrojovka.*

Zehner: Metallwarenfabrik Emil Zehner, Suhl; a maker of 6.35mm vest pistols sold under the "Zehna" tradename between WWI and the late1920s. During the latter half of WWII, the firm made MP44 and other small arms components.

ZEHNA: Tradename of E. Zehner, Suhl; a maker of 6.35mm vest pistols.

Zeilter: Ludwig Zeilter, Wien; manufacturer of edged weapons during the Nazi era.

Zeiss: Carl Zeiss, Jena; a longtime optics manufacturer. Founded by Carl Zeiss in 1846, the firm originally concentrated on manufacturing precision microscopes. By 1875 the firm had sixty employees and over three hundred workers at the time of Zeiss' death in 1888. Zeiss's apprentice Ernst Abbe and Zeiss's son Roderich then took over the company and expanded its operations dramatically. During WWI, the Zeiss firm supplied the Kaiser's military with binoculars, telescopes and other precision optics. After the war, Zeiss improved and aggressively marketed microscopes, cameras, binoculars and telescopic rifle sights for military and sporting rifles. In 1927 Zeiss acquired the majority of shares in the Emil Busch, Rathenow firm. Then in 1928, the concern went even further and boldly acquired a controlling interest in the Hensoldt, Wetzlar firm. Subsequently, Zeiss prospered under the National Socialist regime and was the nation's single largest supplier of optics to the Wehrmacht. Zeiss used a number of proprietary tradenames and trademarks between 1900-1940, such as Deltar, Tessar, Zeil-Dialyt, Zielklein, Zieljagd, Zielmar, Zielmuti, Zielsechs, Zielvier and others. The firm also had subsidiaries such as Zeiss-Contessa and Zeiss-Ikon, and it provided supervision to foreign optical firms operating under German occupation. At the end of the war, the Jena facilities and those of its Dresden subsidiary fell within the Soviet Zone of Occupation.

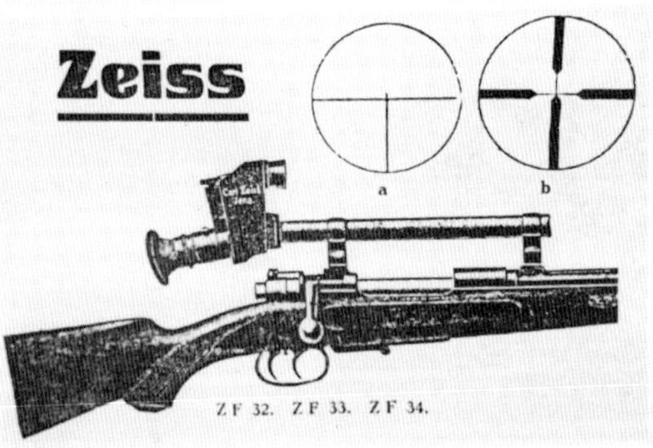

Fig. 182: An early Zeiss advertisement illustrating its ZF32, ZF33, and ZF34 rifle telescopes. Zeiss used a number of different trademarks during the firm's history. This early trademark bears a striking similarity to the one the corporation uses currently.

Between 1945-1947, much of Zeiss' machinery, tooling and inventory were removed to the USSR as war reparations. Carl Zeiss AG, was later reestablished in Oberkochen, West Germany. The former Zeiss facilities located in the German Democratic Republic were merged with the remains of several smaller firms into VEB Carl Zeiss, Jena. The two Zeiss firms then set about battling over trademarks, market share and reputations over several subsequent decades. In the early 1950s, Zeiss also acquired a sizeable share of the Voigtländer firm, and then subsequently absorbed it by the mid-1960s. See also *Voigtländer*, *Hensoldt*, *Busch* and *VEB Carl Zeiss*.

Fig. 183: An advertisement for the famous Zeiss "Zielsechs" scope, ca. 1936. Zeiss had and still has one of the most recognizable trademarks in the global optics industry.

Zeiss: VEB Carl Zeiss, Jena; the post-WWII German Democratic Republic (GDR/DDR) incarnation of Zeiss, reestablished from what remained of the Jena facilities and the assets of several smaller optics manufacturers. As part of war reparations, the Soviets dismantled and shipped to the USSR much of what was left of Zeiss, Jena. What tooling, machinery, and facilities that remained were put back into operation on a small scale in 1947. The first products produced were binoculars, likely made up from left-over wartime components. In 1948, the "People's Enterprise" VEB Carl Zeiss, Jena was formally organized by the East German government. Between the end of WWII and about 1952, there were actually hopes of a post-war German reunification, and thus, the two Zeiss companies cooperated to some extent. Production of microscopes, telescopes, cameras, and gradually lenses ensued. Reunification hopes were dashed however as the Cold War accelerated and the communist East German government increased its control over all businesses in the DDR. Thereafter, the two Zeiss firms set about battling over trademarks, market share and reputations over several subsequent decades. VEB Zeiss was actually one of the few companies in the former DDR that maintained a world-wide reputation for quality and design throughout the Cold War and was a major source of hard currency for the government. By the 1980s,

Fig. 184: Wartime press photo showing a Heer NCO on maneuvers with a pair of 6x30 binoculars made by and trademarked by Zeiss, Jena.

Fig. 185: The destroyed Carl Zeiss, Jena facility just after the close of the Second World War. *US National Archives.*

VEB Carl Zeiss (Jena) and VEB Zeiss Ikon (Dresden) were made up of some 25 smaller companies in the DDR, and employed 70,000 skilled workers.

ZENIT: Tradename used by Moritz & Gerstenberger Waffenfabrik (EM-GE), Zella-Mehlis to market air guns prior to WWII. See also *EM-GE.*

Zico: See *Zimmermann & Co.*

Zi-Di: Tradename used by Heinrich Diem Gewehrfabrik, Benshausen to market .22 caliber target and training rifles, circa 1920-1930s.

Ziel-Dialyt: Tradename used by Hensoldt & Söhne, Mechanisch-Optische Werke AG, Inh. Dr. Hensoldt, Wetzlar to market 2 ¾- 8 power rifle telescopes in the 1930-1940s.

Zieleins: Tradename used by Carl Zeiss, Jena for low-power rifle telescopes, 1930-1940s. See also *Zeiss.*

Zielklein: Tradename used by Carl Zeiss, Jena for low-power rifle telescopes, 1930-1940s. See also *Zeiss.*

Zieljagd: Tradename used by Hensoldt & Söhne, Mechanisch-Optische Werke AG, Inh. Dr. Hensoldt, Wetzlar to market rifle telescopes.

Zielmar: Tradename used by Carl Zeiss, Jena for low-power rifle telescopes, 1930-1940s. See also *Zeiss.*

Zielmuti: Tradename used by Carl Zeiss, Jena for variable 1-4 power rifle telescopes, 1930-1940s. See also *Zeiss.*

Zielmuti: Tradename used by Carl Zeiss, Jena for variable 1-6 power rifle telescopes, 1930-1940s. See also *Zeiss.*

Zielsechs: Tradename used by Carl Zeiss, Jena for 6-power rifle telescopes, 1930-1940s. See also *Zeiss*.

Zielvier: Tradename used by Carl Zeiss, Jena for 4-power rifle telescopes, 1930-1940s. See also *Zeiss*.

Zimmermann: Zimmermann & Co., "Zico", Preß- u. Stanzwerk, Remscheid; a cutting tool manufacturer founded in 1866. The firm used the "Zico" trademark, first registered in 1932. During WWII, Zimmermann made Pak36 wrenches and other tools for the Wehrmacht.

Appendix I: Manufacturers by Observed Trademarks/Logos

Trademark	Manufacturer
aba	"Aba-Metall", Allgemeine Berliner Automaten- u. Metallwarenfabrik GmbH, Berlin
	Adamy Gebr., Suhl
	"ADLER FLINTE" Anton Ferdinand Korinski, Erlangen
	Adlerwaffenwerke, Engelbrecht & Wolff, Zella St. Blasii
AEG	"AEG" Allgemeine Elektrizitäts-Gesellschaft, Berlin
	"AEG" Allgemeine Elektrizitäts-Gesellschaft, Berlin [early trademark variant]
AEG	"AEG" Allgemeine Elektrizitäts-Gesellschaft, Berlin [post-WWII]
AESCULAP	"Aesculap" Aktiengesellschaft für Feinmechanik, vormals Jetter & Scheerer, Tuttlingen
	"Aesculap" Aktiengesellschaft für Feinmechanik, vormals Jetter & Scheerer, Tuttlingen [variant]
Agfa	AGFA; I.G. Farben-Industrie AG, Werk "Agfa", München

	"Ajack" A. Jackenkroll, Optische Anstalt GmbH, Berlin [early]
AJACK	"AJACK" A. Jackenkroll, Optische Anstalt GmbH, Berlin [variant]
Akah	Albrecht Kind, Berlin-Nürnberg
SOLINGEN	"Alcoso" GmbH, Alexander Coppel, Stahlwaren- u. Waffenfabrik, Solingen
A C S	"Alcoso" GmbH, Alexander Coppel, Stahlwaren- u. Waffenfabrik, Solingen [variant]
A C S	"Alcoso" GmbH, Alexander Coppel, Stahlwaren- u. Waffenfabrik, Solingen [variant]
	"ALFA" Adolf Frank, Hamburg
ALBERT MEBUS AMSO SOLINGEN	"AMSO" Albert Mebus, Solingen
AMBERG	"Amberg" Königliche Gewehrfabrik Amberg, Bayern
	Anker-Werke, Bielefeld [early]
ANKER WERKE BIELEFELD	Anker-Werke, Bielefeld [variant]
	Anker-Werke, Bielefeld [later]

	J.G. Anschütz, GmbH & Co. KG, Ulm/Do [post-WWII]
	Gustav Appel Maschinenfabrik, Berlin-Spandau
	Argenta-Gesellschaft, Solingen
	"ARMESO" Arthur Melcher, Solingen-Merscheid
	"ARMINIUS" Friedrich Pickert, Zella-Mehlis
	Askania-Werke AG, Berlin
	Assmann & Söhne, Lüdenscheid
	"ASSO" Arthur Schüttelhöfer & Co., Solingen
	Astra-Unceta y Cia. S.A., Eibar (later Guernica), Spain
	Astra-Werke, Chemnitz
	Auer-Gesellschaft AG, Guben
	Auer-Gesellschaft AG, Berlin [post-WWII]

	"AW" Feinwerkbau Westinger & Altenburger, Oberndorf [post-WWII]
	"AW" Feinwerkbau Westinger & Altenburger, Oberndorf [post-WWII]
	"B" Rheinmetall-Borsig AG, Sömmerda [variant, transition]
	Braun & Bloem
	F.W. Backhaus, Solingen
	Julius Bahrl, Solingen
	Fa. Walter Bahrl, Solingen
	Richard Balke & Söhne, Solingen
	Fritz Barthelmess, Muggendorf
	"BAYARD" Anciens Etablissements Pieper, Herstal, Belgium
	Ludwig Bayerl, Büchsenmacher, Regensberg
	Eduard Becker, Solingen

Trademark	Company
	Gebr. Becker, Solingen
	Becker und Holländer, Suhl
GEBR. BELL SOLINGEN-GR	Gebr. Bell, Solingen
	Gebr. Bell, Solingen
CARL BENDER SOLINGEN	Carl Bender, Solingen
	Fabbrica d'Armi Pietro Beretta, Gardone VT (Brescia), Italy
	Fabbrica d'Armi Pietro Beretta, Gardone VT (Brescia), Italy [post-WWII]
BERGMANN	Theodor Bergmann Maschinen u. Waffenfabrik, Suhl
	Theodor Bergmann Maschinen u. Waffenfabrik, Suhl [variant]
B	Theodor Bergmann Maschinen u. Waffenfabrik, Suhl [variant]
Th B Suhl	Theodor Bergmann Maschinen u. Waffenfabrik, Suhl [variant]
BLM	Berlin-Lübecker Maschinenfabrik, Lübeck

	Berlin-Suhler Waffen- und Fahrzeugwerke, Weimar
	Berlin-Suhler Waffen- und Fahrzeugwerke, Weimar [variant]
	Gebr. Berns, Solingen
	Hugo Berns, Solingen-Ohligs
	C. Reinhard Bertram & Sohn, Solingen
	C. Beuttenmüller & Cie., Metallwarenfabrik, Bretten/Baden
	August Bickel, Steinbach-Hallenberg
	Gebr. Bing, Nürnberg
	"BISMARCK" August Müller, Solingen
	Gustav Bittner Waffenfabrik, Weipert/Erzgebirge (Vejprty), Czech [early]
	Gustav Bittner, Waffenfabrik, Weipert/Erzgebirge (Vejprty), Czech
	Blohm & Voss, Schiffswerft u. Maschinenfabrik, Hamburg

Mark	Entry
	Julius Bodenstein, Steinbach
	"Dr. B" Dr.-Ing. Boehme & Co., Werksleitung Lüdenscheid, Werk Minden
	"BÖHLERSTAHL" Böhler & Co. AG, Edelstahlwerke, Kapfenberg
	"BÖHLERSTAHL" Böhler & Co. AG, Edelstahlwerke, Kapfenberg [variant]
	"BÖHLERSTAHL" Böhler & Co. AG, Edelstahlwerke, Kapfenberg [variant]
	"BÖHLERSTAHL" Böhler & Co. AG, Edelstahlwerke, Kapfenberg [variant]
	"BÖHLERSTAHL" Böhler & Co. AG, Edelstahlwerke, Kapfenberg [variant]
	"BÖHLERSTAHL" Böhler & Co. AG, Edelstahlwerke, Kapfenberg [variant]
	Gebrüder Böhme Nachf., Brotterode
	Heinrich Böker, Baumwerk, Solingen
	Karl Böker, Solingen
	Bolte & Anschutz Jagdwaffenfabrik, Zella-Mehlis

	"Bombe" Deutsche Waffen- und Munitionsfabriken AG, Berlin and Karlsruhe
	Böntgen & Sabin, Solingen
	Böntgen & Sabin, Solingen
	Gebrüder Born, Solingen
	Gebrüder Born, Solingen [variant]
	August Borsig, Berlin-Tegel [early]
	"BORSIG" Rheinmetall-Borsig AG, Sömmerda [later]
	Brandenburger Fahrrad- u. Motorrad-Werke "Excelsior" GmbH, Brandenburg
	Wilhelm Brenneke Gewehr u. Geschossfabrik, Leipzig
	Ernst Bruckmann, Solingen-Ohligs
	Ernst Friedrich Büchel, Suhl
	Rudolf Bückel, Solingen-Merscheid

Mark	Company
	Karl Burgsmüller, Berlin
	Emil Busch A.G., Rathenow
	Emil Busch A.G., Rathenow [variant]
	Emil Busch A.G., Rathenow [variant]
	Ernst Busch, Solingen
	Busch-Jaeger Lüdenscheider Metallwerke AG, Lüdenscheid
	"BVG" Bochumer Verein für Gußstahlfabrikation AG, Bochum
	"Cewe" Carl Walther Waffenfabrik, Zella-Mehlis, Thüringen [molded products]
	Gebr. Christians, Solingen
	Christianswerk, Solingen
	"CHROMOLIT" Anton Wingen, Solingen
	Edward Cleff, Solingen

ČSZ	"CSZ" Česká Zbrojovka, Czechoslovakia [pre-WWII]
CURDTSNACHF SOLINGEN	Curdtsnachf, Solingen
Z	"CZ" Česká Zbrojovka, Czechoslovakia [renamed Böhmische Waffenfabrik AG, Prag under German occupation during WWII]
	Deutsche-Werke AG, Erfurt
	Deutsche-Werke AG, Erfurt [variant]
D	"D" Gustav Genschow & Co. AG, Berlin-Treptow
DÖHLENSTAHL	"D" Sächsische Gußstahlwerke Döhlen AG, Freital [after 1927]
D&C	"D&C" Dransfeld & Co., Spezialfabrik für Beschlagteile aller Art, Menden
HOFFNUNGSWERK ERNST DIRLAM SOLINGEN	Ernst Dirlam Hoffnungswerk, Solingen
	Dansk Industri Syndikat AG "Madsen", Kopenhagen
DISA	Dansk Industri Syndikat AG "Madsen", Kopenhagen [post-WWII trademark]
DANZIG	Königlich-Preussische Gewehrfabrik Danzig, Danzig

Mark	Company
DECKER.	Walter Decker, Zella-St. Blasii
DEMAG	"DEMAG" Deutsche Maschinenfabrik AG, Duisberg
Diamant	"Diamant" Elite-Diamantwerk, Siegmar-Schönau bei Chemnitz
	"DIANA" Dianawerk Mayer & Grammelspacher, Rastatt
F.DICK	Friedrich Dick, Esslingen
F.D.	Friedrich Dick, Esslingen [variant]
Diem	"Diem" Heinrich Diem Gewehrfabrik, Benshausen bei Suhl
L.O.DIETRICH	L.O. Dietrich, Altenburg [ca. 1900]
	L.O. Dietrich, Altenburg [ca. WWII]
DINOX	"DINOX" W. Bischoff, Braunschweig [post-WWII]
DIRLAM & SÖHNE SOLINGEN	Dirlam & Söhne, Solingen
J.E. DITTERT & Co	J. E. Dittert & Co

Trademark/Logo	Manufacturer
GDC	G.C. Dornheim AG, Waffenfabrik, Suhl
ALBERT DORSCHEL · SOLINGEN	Albert Dorschel, Solingen
RICH. DREES & SOHN SOLINGEN	Richard Drees & Sohn, Solingen
DSt	"DSt" Sächsische Gußstahlwerke Döhlen AG, Freital [before 1927]
DUO	"DUO" Kohout & Spol.(SPK), Kdyne, Czechoslovakia
	Dürkopp Werke, Bielefeld [ca. 1900]
DÜRKOPP	Dürkopp Werke, Bielefeld
DÜRKOPP	Dürkopp Werke, Bielefeld [variant]
DÜRKOPP	Dürkopp Werke, Bielefeld [post-WWII]
Durofol	Durofol; unconfirmed- trademark was last registered under Hans Joachim Gerling, Solingen-Ohligs (post-WWII)
DWM	"DWM" Deutsche Waffen- und Munitionsfabriken AG, Berlin [early]
DWM	"DWM" Deutsche Waffen- und Munitionsfabriken AG, Berlin and Karlsruhe

DWM	"DWM" Deutsche Waffen- und Munitionsfabriken AG, Berlin and Karlsruhe [later]
	Dynamit-Actien-Gesellschaft vormals Alfred Nobel & Co., Abteilung Celluloid-und-Kunststoff-Fabrik, Werk Troisdorf (Bez. Köln)
	Dynamit AG, Werk Troisdorf, Troisdorf [post-war variant]
	Deutsche Edelstahlweke AG, Remscheid
PAUL EBEL SOLINGEN	Paul Ebel, Solingen
EBONSMANN SOLINGEN	Ebonsmann, Solingen
ECKO	"ECKO" Emil Eckold Waffenfabrik, Suhl
EE	"ERMA" Erfurter Maschinenfabrik Berthold Geipel GmbH, Erfurt
EICKELNBERG & MACK SOLINGEN	Eickelnberg & Mack, Solingen
EICKHORN SOLINGEN	Waffenfabrik Carl Eickhorn, Solingen [early variant]
ORIGINAL Eickhorn	Waffenfabrik Carl Eickhorn, Solingen [later variant]
ORIGINAL Eickhorn SOLINGEN	Waffenfabrik Carl Eickhorn, Solingen [modern]

Logo	Manufacturer
Elfa	"Elfa" Elsterwerdaer Fahrradfabrik, E.W. Reichenbach GmbH, Elsterwerda-Biehla
Elite-Diamant	"Elite-Diamant" Elite-Diamantwerk, Siegmar-Schönau bei Chemnitz
EM-GE	"EM-GE" Moritz & Gerstenberger Waffenfabrik (EM-GE), Zella-Mehlis
ERFURT	Königlich-Preussische Gewehrfabrik Erfurt, Erfurt, Thüringen
C. EPPENSTEIN-SÖHNE SOLINGEN	C. Eppenstein & Söhne, Solingen
	"ERMA" Erfurter Maschinenfabrik Berthold Geipel GmbH, Erfurt [early]
ERMA	"ERMA" Erfurter Maschinenfabrik Berthold Geipel GmbH, Erfurt
ERMA ERFURT	"ERMA" Erfurter Maschinenfabrik Berthold Geipel GmbH, Erfurt [variant]
FABRIKAT ERMA ERFURT	"ERMA" Erfurter Maschinenfabrik Berthold Geipel GmbH, Erfurt [variant]
ERMA	"ERMA" Erfurter Maschinenfabrik Berthold Geipel GmbH [post-WWII]
Esha	"Esha" E. Schmidt & Habermann Jägdgewehrfabrik, Suhl
A E S	Arthur Everz, Solingen

EZ	"EZ" Zehna Metallwarenfabrik Emil Zehner, Suhl
F	"F" Feinwekbau Westinger & Altenburger, Oberndorf [post-WWII]
FALKE	Falkewerke, Wennigsen [post-WWII]
Feinprüf	"Feinprüf" Feinprüf Feinmeß- und Prüfgeräte GmbH, Göttingen
FÉG	"FEG" Femaru-Fegyveres Gépgyár R.T., Budapest, Hungary [modern]
F F G	"FFG" Femaru-Fegyveres Gépgyár R.T., Budapest, Hungary
G.FELIX GLORIAWERK SOLINGEN	Gustav Felix, Gloriawerk, Solingen
FICHTEL & SACHS SCHWEINFURT	Fichtel & Sachs, Schweinfurt
	Fichtel & Sachs "Torpedo", Schweinfurt [later variant]
FISCHERWERK SOLINGEN	Fischerwerk, Solingen
FB	"FB" Fabryka Broni W. Radomiu, Radom, Poland
FN	"FN" Fabrique Nationale D'Armes de Guerre, Herstal Belgium [early]

Logo	Manufacturer
	"FN" Fabrique Nationale D'Armes de Guerre, Herstal Belgium
Frankonia	Frankonia Jagd Hofmann GmbH & Co. KG, Rottendorf [post-WWII]
Fridericus Solingen	Fridericus, Solingen
FROMMIA	"FROMMIA" Ferdinand Fromm Maschinenfabrik, Stuttgart-Bad Cannstatt
	"FS" Femaru-Fegyveres Gépgyár R.T., Budapest, Hungary [Frommer Stop]
R FUESS BERLIN-STEGLITZ	R. Fuess, Berlin-Steglitz
CHR. FUNK 1835	Christoph Funk Jagdgewehrfabrik, Suhl
	Dr. Walter Gerard, Berlin-Charlottenburg
	Gustloff-Werke, Suhl
	Gustloff-Werke, Suhl [variant]
	Gustloff-Werke, Suhl [variant]
	Fabrique d'Armes de Guerre Haute Precision Armand Gavage, Liège, Belgium

GARANTIE	"GARANTIE" Gebrüder Pauh, Solingen
GAZELLE	"GAZELLE" Gebrüder Krusius, Solingen
GECADO	"GECADO"; originally G.C. Dornheim AG, Waffenfabrik, Suhl
GECADO	"GECADO"; originally G.C. Dornheim AG, Waffenfabrik, Suhl [variant]
Geco	"Geco" Gustav Genschow & Co. AG, Berlin-Treptow
Ge co	"Geco" Gustav Genschow & Co. AG, Berlin-Treptow [variant]
	"Geco" Gustav Genschow & Co. AG, Berlin-Treptow [variant]
Geha	"Geha" Gustav Genschow, Gebr. Heym and Gebr. Adamy, others
FRIEDRICH GEIGES SOLINGEN-FOCHE	Friedrich Geigis, Solingen-Foche
Uhlenflucht	Robert Geller, Giessen [probably post-WWII]
ED. GEMBRUCH SOLINGEN	Eduard Gembruch, Solingen
E. GIERLING SOLINGEN	Emil Gierling, Solingen

Logo	Manufacturer
	Robert Giersch, Solingen
	Giesen & Forsthoff, Solingen
	"Goerz" Zeiss-Ikon AG, Optische Geräte, Goerz-Werk, Berlin-Zehlendorf
	"Goerz" Zeiss-Ikon AG, Optische Geräte, Goerz-Werk, Berlin-Zehlendorf [variant]
	"Goerz" Zeiss-Ikon AG, Optische Geräte, Goerz-Werk, Berlin-Zehlendorf [variant]
	"Goerz" Zeiss-Ikon AG, Optische Geräte, Goerz-Werk, Berlin-Zehlendorf [variant]
	Gebrüder Gräfrath, Fabrik feiner Stahlwaren, Solingen-Widdert
	Carl Grah, Solingen-Ohligs
	Ernst Grah, Solingen-Wald
	Gustav Reinhold Grah, Solingen-Foche
	"Gränriso" Gebrüder Gräfrath, Fabrik feiner Stahlwaren, Solingen-Widdert
	Greifelt & Co. Gewehrfabrik, Suhl

	Greifelt & Co. Gewehrfabrik, Suhl [variant]
	Ludwig Groten, Solingen
	"Gwf.A" Gewehrfabrik Amberg, Bäyern
	Carl Haas, Solingen-Wild
	R. Haastert & Bull, Solingen-Wald
	"Habicht" D. Swarovski, Glasfabrik u. Tyrolit-Schleifmittelwerke, Tyrol, Austria
	HACO-Werk, Ernst Mandewirth, Berlin
	"Hack-Werke" Josef Hack GmbH, Steyr/Ober Donau
	C.G. Haenel Waffen- und Fahrradfabrik AG, Suhl
	Eugen Haering, Solingen
	Hermann Hahn, Solingen-Wald
	Gustav Häker; Solingen

	Carl Halbach, Solingen
	Gebr. Halbach, Weyer b.Ohligs
	Wilhelm Halbach, Solingen
	Hammesfahr & Cie, Solingen
	Gottlieb Hammesfahr Stahlwarenfabrik AG, Solingen-Foch
	Hartkopf & Co, Solingen
HARTMANN WETZLAR	Hartmann-Optik Wetzlar, Werkstätten für Feinoptik und Feinmechanik Inh. Karl Hartmann, Wetzlar [believed to be post-WWII]
HASAG	"HASAG" Hugo Schneider AG, Leipzig
	"HAWIG" Hauswirtschafts-Maschinen-Gesellschaft mbH, Berlin
HDK	Hackethal-Draht- u. Kabelwerke AG, Hannover
	Carl Heindelberg; Solingen
	F. & A. Helbig, Steinbach, bei Schleusingen

	"K. Helfricht" A. Krausser, Zella-Mehlis
	Gebr.Heller, Marienthal bei Schweina
	Gebr.Heller, Schmalkalden
	Henckels AG, Zwillingswerk, Schneidwarenfabrik u. Stahlwerk, Solingen
	Henschel & Sohn GmbH, Kassel
	Henschel & Sohn GmbH, Kassel [variant]
	Henschel & Sohn GmbH, Kassel [variant]
	Hensoldt & Söhne, Mechanisch-Optische Werke AG, Inh. Dr. Hensoldt, Wetzlar and Herborn [early]
HENSOLDT WETZLAR	Hensoldt & Söhne, Mechanisch-Optische Werke AG, Inh. Dr. Hensoldt, Wetzlar and Herborn [later]
HERBECK & MEYER SOLINGEN-HÖHSCHEID	Herbeck & Meyer, Solingen-Höhscheid
	Herbertz & Meurer, Solingen-GR
	Herbertz & Meurer, Solingen-GR

Trademark	Manufacturer
HERDER & ENGELS ★SOLINGEN-OHLIGS★	Herder & Engels, Solingen-Ohligs
F. HERDER A.S. SOLINGEN	Friedrich A. Herder und Sohn, Solingen
HERDER	Friedrich A. Herder und Sohn, Solingen [variant]
	Friedrich A. Herder und Sohn, Solingen [variant]
H. HERDER + SOLINGEN +	H. Herder, Solingen
	H. Herder, Solingen
RICH. ABR. HERDER · SOLINGEN ·	Richard A.Herder, Solingen
H RICH. ABR. HERDER	Richard A.Herder, Solingen [variant]
RICH.A.HERDER SOLINGEN	Richard A.Herder, Solingen [post-WWII]
	Robert Herder, Solingen
DIOGENES	Herder & Sohn, Solingen
"Herold"	"Herold" Franz Jäger & Company, Suhl.

	Hertel & Reuss, Optik u. Feinmechanik, Kassel
	Herz, Berlin
	Franz W. Heym, Suhl
	Friedrich Wilh. Heym GmbH & Co. KG, Münnerstadt [post-WWII]
	"H.H.K." Heinrich H. Klüssendorf, Werkzeuge, Berlin Spandau
	Hirtenberger Patronen- und Zündhütchenfabrik, Hirtenberg
	Hahn & Kolb, Werkzeugmaschinen-Fabrik, Stuttgart
	"HO" Deutsche-Werke AG, Erfurt [Heinrich Ortgies]
	Matthias Hohner AG, Harmonikafabrik, Trossingen
	F.W. Höller, Solingen
	Hommelwerke GmbH, Mannheim-Käfertal
	Curt Hoppe, Solingen

Logo	Manufacturer
WILHELM HOPPE SOLINGEN	Wilhelm Hoppe, Inh. Helmet Giesen, Solingen
E&F HÖRSTER SOLINGEN	E&F Hörster Stahl- u. Metallwarenfabrik, Solingen
HHS	E&F Hörster Stahl- u. Metallwarenfabrik, Solingen [variant]
HR	"HR" Hagener Gußstahlwerke, Remy & Co. GmbH, Hagen
HS	"HS" Herbert Schmidt Waffenfabrik, Zella-Mehlis [modern]
Hubertus	"Hubertus" Imman. Meffert Gewehrfabrik, Berlin
HUBERTUS.	"HUBERTUS" Imman. Meffert Gewehrfabrik, Berlin [variant]
„Hubertus"	"Hubertus" Imman. Meffert Gewehrfabrik, Berlin [variant]
HW	"HW" Hommelwerke GmbH, Mannheim-Käfertal [unconfirmed]
HWM	"HWM" Hermann Weihrauch Gewehr und Fahrradteile Fabrik, Zella-Mehlis
HWZ	"HWZ" Hermann Weihrauch Gewehr und Fahrradteile Fabrik, Zella-Mehlis [variant]
WEIHRAUCH HWZ	"HWZ" Hermann Weihrauch Gewehr und Fahrradteile Fabrik, Zella-Mehlis [variant]

	"HWZ" Hermann Weihrauch Gewehr und Fahrradteile Fabrik, Zella-Mehlis [variant]
	"Ideal" Feinbau Maschinen GmbH, Winterbach-Stuttgart
	"iG" Industrie Gesellschaft Farben
	Industriewerke Karlsruhe AG, Karlsruhe [post-WWII]
	Jacobs & Co., Solingen-Gräfrath
	Rudolf C. Jacobs, Solingen-Gräfrath
	"JGA" J.G. Anschütz, Germaniawaffenfabrik, Zella-Mehlis, Thüringen
	Jenoptik, Jena [post-WWII optical firm]
	Jenoptik, Jena [post-WWII optical firm; variant]
	F. W. Jordan, Solingen
	Justinus Werk, Solingen
	Justinus Werk, Solingen [post-WWII]

Logo	Manufacturer
Kaba Spezial	"Kaba" originally Karl Bauer & Co., Berlin; later Gustav Genschow, Berlin
KKAHLES WIEN	Kahles, Karl, Optiker, Wein
EMIL KAISER & Co SOLINGEN	Emil Kaiser und Co., Solingen
KARL ROB. KALDENBACH KKR SOLINGEN-GR	Karl Robert Kaldenbach, Solingen
K M F	"KMF" Kallenbach, Meyer & Franke, Metallwarenfabrik, Luckenwalde
KASTO	Kasto Maschinenbau GmbH & Co. KG., Achen
KAUFMANN & SÖHNE SOLINGEN	Heinrich Kaufmann & Söhne, Solingen
K55K	Heinrich Kaufmann & Söhne, Solingen [variant]
K55K	Heinrich Kaufmann & Söhne, Solingen [variant]
ERNST KEMPER SOLINGEN	Ernst Kemper, Solingen
Akah	Albrecht Kind, Berlin-Nürnberg
	Albrecht Kind, Berlin-Nürnberg [variant]

A. K.	Albrecht Kind, Berlin-Nürnberg [variant]
„SCHUTZMANN“	Albrecht Kind, Berlin-Nürnberg [variant]
MARKE EICHE	Albrecht Kind, Berlin-Nürnberg [variant]
Akah	Albrecht Kind, Berlin-Nürnberg [variant]
FA. KIRSCHBAUM &Co SOLINGEN	F.A & Co Kirschbaum, Solingen
	Robert Klass Messerfabrik, Solingen-Ohligs [pre-1900]
	Robert Klass Messerfabrik, Solingen-Ohligs
ROBERT KLASS SOLINGEN	Robert Klass Messerfabrik, Solingen-Ohligs
RK	Robert Klass Messerfabrik, Solingen-Ohligs [post-WWII]
KLITTERMANN & MOOG GmbH HAAN b/SOLINGEN	Klittermann & Moog GmbH, Haan b/Solingen
AUG. KNECHT SOLINGEN	August Knecht, Solingen
K. & Co SOLINGEN	E. Knecht & Co., Solingen

KNOCH	Nähmaschinenfabrik Adolf Knoch AG, Saalfeld/Saale
KNOCH	Nähmaschinenfabrik Adolf Knoch AG, Saalfeld/Saale [variant]
K	Knorr-Bremse AG, Berlin Nähmaschinenfabrik Adolf Knoch AG, Saalfeld/Saale
WILH. KOBER & CO. SUHL GEGR. 1875	Wilhelm Kober & Co., Suhl
KOBRA	"KOBRA" Walter Diefke, Suhl
	Paul Kohl, Solingen
	F. Koeller; Solingen-Ohligs
FK F R. v.D KOHLEN SOLINGEN	Friedrich v.d Kohlen; Solingen
G.L. KÖLLER NACHE SOLINGEN	Gustav L. Köller, Solingen
HUGO KÖLLER SOLINGEN	Hugo Köller, Solingen
H. KOLPING SOLINGEN	H. Kolping, Solingen
Th. K.	Theodor Kommer, Zella-Mehlis

	Theodor Kommer, Zella-Mehlis [later variant]
	Hermann Konejung AG, Solingen
	Hermann Konejung AG, Solingen
	Peter Daniel Krebs, Solingen
	Peter Daniel Krebs, Solingen
	Carl Julius Krebbs, Solingen
	Georg Krempf, Aktiengesellschaft für Optik u. Mechanik, Wetzlar
	"Krico" Kriegskorte & Co. GmbH, Stuttgart
	Wilhelm Krieger, Solingen
	Heinrich Krieghoff Waffenfabrik, Suhl
	Heinrich Krieghoff Waffenfabrik, Suhl [variant]
	Heinrich Krieghoff Waffenfabrik, Suhl [variant]

Heinrich Krom, München

W. Krombach, Wetzlar

"KRONECK" Ernst Erich Witte, Solingen

"KRONECK" Ernst Erich Witte, Solingen [variant]

Gebr. Krumm, Solingen

Fried. Krupp AG, Essen

Fried. Krupp AG, Essen [variant]

Fried. Krupp AG, Essen [modern]

Gebrüder Krusius, Solingen

Kuhagen, Lüchow [post-WWII]

Carl Friedrich Kührt KG, Zella-Mehlis

August Küllenberg, Solingen-Gräfrath

	Lang: Federn-, Draht- u. Metallwarenfabrik Rudolf Lang, Brandenburg/Havel
FL	Fr. Langenhan Gewehr- und Fahrradfabrik, Zella-Mehlis
H&F. LAUTERJUNG W SOLINGEN	H. & F. Lauterjung, Solingen
Leica	Ernst Leitz, GmbH., Optische Werke, Wetzlar
Leico	Ernst Leitz, GmbH., Optische Werke, Wetzlar
Leitz	Ernst Leitz, GmbH., Optische Werke, Wetzlar
E.LEITZ WETZLAR	Ernst Leitz, GmbH., Optische Werke, Wetzlar [variant]
L&F	"L&F" Linden & Funke KG, Iserlohn
LEUCO	"LEUCO" Johann Leupold, Bayreuth
	Johann Leupold, Bayreuth
LIGNOSE	Lignose AG, Berlin, Abteilung Suhl
C&R. LINDER C&RL SOLINGEN-WEYER	C & R Linder, Solingen-Weyer

	Hugo Linder Deltawerk, Solingen
	Hugo Linder C. W. Sohn, Solingen-Weyer
	Hermann Linder Söhne, Solingen
	Otto Linder Solingen-Merscheid
	Ludwig Loewe & Co., Berlin
	Loewe Fabriken, Berlin [ca. WWII]
	Loffe, Solingen
	Lohmann-Werke AG, Fahrrad- u. Motorensättel-Fabrik, Bielefeld
	Louper, Solingen
	P.D. Lüneschloss, Solingen
	P.D. Lüneschloss, Solingen [variant]
	Peter Lungstrass, Solingen-Ohligs

Mark	Company
C. LÜTTERS & CIE LÖWENWERK	C. Lütters & Cie Löwenwerk, Solingen
E. LÜTTGES & Co SOLINGEN	E. Lüttges & Co., Solingen
L LUX X	Fa. Emil Lux, Remscheid
L W	"LW" Lohmann-Werke AG, Fahrrad- u. Motorensättel-Fabrik, Bielefeld
MAB	"MAB" Manufacture d' Armes des Bayonne, Bayonne, France
Macero HENKEL & MÜLLER SOLINGEN-O.	"Macero" Henkel & Müller, Solingen
Gegründet 1882	Bruno Mädler, Berlin
	"MAGET" Maschinen-u Gerätebau GmbH, Berlin-Tegal
MAIHAK	H. Maihak AG, Fabrik für techn. Meßinstrumente u. feinmech. Geräte, Hamburg
M A	August Malsch Fr Sohn, Steinbach, Kr. Meinigen
DAVID MALSCH DM STEINBACH	David Malsch, Steinbach Kr. Meinigen
Karl Malsch Gus & Sohn Steinbach Kr. Mein.	Karl Malsch, Steinbach, Kr. Meinigen

	Karl Malsch Spitzer, Steinbach, Kr. Meinigen
	Malsch & Ambronn, Steinbach
	"M.A.N." Maschinenfabrik Augsburg-Nürnberg AG
	Fritz Mann, Suhl
	Fritz Mann, Suhl [variant]
	"MANURHIN" Mre. de Machines du Haut Rhin, France [post-WWII]
	"MANURHIN" Mre. de Machines du Haut Rhin, France [post-WWII]
	Richard Marholdt & Son, Tiroler Waffenfabrik Peterlongo, Innsbruck, Austria
	"MaRie" Mayer & Riem, Arnsburg
	"MARS" Kohout and Spol, Kdyne, Czech
	"MARS" Kohout and Spol, Kdyne, Czech [variant]
	"MARS" Kohout and Spol, Kdyne, Czech [variant]

	"MAS" Manufacture Nationale d'Armes de Saint Etienne (MAS), St. Etienne, France
	"MASSI" Mass-Industrie: Mass-Industrie GmbH, Meßwerkzeugfabrik, Werdau
	Alphons Mauser GmbH, Köln-Ehrenfeld
	Gebr. Mauser & Cie, Oberndorf a.N.
	Mauser-Werke AG, Oberndorf a.N. [early]
	Mauser-Werke AG, Oberndorf a.N. [later variant]
	Mauser-Werke AG, Oberndorf a.N. [later variant]
	Mauser-Werke AG, Oberndorf a.N. [variant post-WWII trademark]
	May und vom Hau, Solingen-Ohligs
	C & W. Meinel-Scholer, Metallwaren, Klingenthal
	Carl August Meis GmbH, Solingen
	Melzer & Feller, Zella-Mehlis

	Waffenfabrik August Menz, Suhl
	Waffenfabrik August Menz, Suhl [variant]
	Waffenfabrik August Menz, Suhl [variant]
	Waffenfabrik August Menz, Suhl [used on Lilliput pistols]
	Meopta, Prerov, Czechoslovakia [post-WWII]
	Meopta, Prerov, Czechoslovakia [variant, post-WWII]
	Gebr. Merkel, Suhl
	Gebr. Merkel, Suhl [variant]
	Gebr. Merkel, Suhl [post-WWII]
	August Merten Ww, Solingen-Grafrath
	Merz-Werke, Gebr. Merz, Metallwarenfabrik, Frankfurt-Rödelheim
	J. D. Moeller, Optische Werke GmbH, Wedel

	J. D. Moeller, Optische Werke GmbH, Wedel [post-WWII]
	Gottfried Müller, Herges-Vogtel
	Müller & Schmidt Pfeilringwerk, Solingen
	Josef Münch, Brotterod
	"MUNDLOS" Mundlos Aktien Nähmaschinenfabrik, Berlin-Magdeburg [ca. 1900]
	"NAHMATAG" Nähmaschinenteile-AG, Spezialfabriken f. Nähmaschinen-schiffchen und -teile aller Art, Radebeul, bei Dresden
	Gustav Neeff, Solingen
	Neidhardt & Schmidt Co., Brotterode
	"NEPTUN" Küpper & Oertling, Solingen
	Ferdinand Neuhaus, Solingen
	Neunkircher Eisenwerk AG (vormals Gebr. Stumm), Homburg (Saar)
	B. Nickel, Marburg [thought to be post-WWII]

B.NICKEL MARBURG	B. Nickel, Marburg [thought to be post-WWII]
Nico	Nico-Pyrotechnik Hanns-Jürgen Diederichs GmbH & Co., Trittau [presumably]
NICORRO	"Nicorro" Deutsche Waffen- und Munitionsfabriken AG, Berlin and Karlsruhe
Nitschke	Nitschke, Rathenow
GvN	Gotthilf von Nordheim, Zella-Mehlis
OCULUS	"Oculus" Spezialfabrik opthalmologisher Instrumente, Berlin
Œ WG	Österreich. Waffenfabriks Ges. Steyr [an early marking for what eventually became Steyr-Daimler-Puch AG, Steyr]
JULIUS OHLIGER SOLINGEN	Ohliger Julius, Solingen
OIGEE BERLIN	"OIGEE" Optische Anstalt Oigee, Berlin-Schöneberg
Olympia	Olympia Büromaschinenwerke AG, Erfurt
	Olympia Büromaschinenwerke AG, Erfurt [variant]
OMNIPOL	Omnipol, Prague (Praha) [post-WWII]

	"OP" Optikotechna/Meopta, Prerau [ca. WWII]
	"OP" Optikotechna/Meopta, Prerau [post-WWII]
	"OTHELLO" Anton Wingen Jr., Solingen
	"ÖWA" Österreichische Werk gws Anstalt, Wien
	"ÖWG" Österreichische Waffenfabriks-Gesellschaft Steyr, Steyr
	Bernhard Paatz, Zella-Mehlis [later variant]
	Bernhard Paatz, Zella-Mehlis [later variant]
	Ernst Pack und Söhne Stahlwarenfabrik, Solingen
	"Parabellum" tradename originally owned by DWM, but currently Rheinmetall Waffe u. Munition GmbH, Ratingen
	Carl Pauls, Solingen
	PEBRA: Paul Braun Präzisionszieherei und Metallwarenfabrik, Esslingen a.N
	"PECAR" P. Köhler, Berlin

	Daniel Peres, Solingen
	Perfectum, Solingen
	Franz Pfannl Waffen- und Patronenfabrik, Krems a/d Donau
	Friedrich Pickert, Zella-Mehlis
	Friedrich Pickert, Zella-Mehlis [variant]
	Friedrich Plücker, Solingen
	Richard Plümacher, Solingen
	Polte AG, Armaturen- u. Maschinenfabriken, Magdeburg [early]
	Polte AG, Armaturen- u. Maschinenfabriken, Magdeburg [later]
	"PRAGA" Zbrojovka Praga, Prague, Czechoslovakia
	Pressstoffwerk Schöppenstedt, Paul Schnake, Schöppenstedt
	Presswerk A.G., Essen

	"Presto" Hagener Gußstahlwerke, Remy & Co. GmbH, Hagen
	"PUMA" Pumawerk, Lauterjung & Sohn GmbH & Co., Solingen
	"PUMA" Pumawerk, Lauterjung & Sohn GmbH & Co., Solingen [variant]
	"R" Rheinisch-Westfalische Sprengstoff Aktiengesellschaft, Karlsruhe
	"R" Ruberg & Renner, Inh. Jos. Ruberg, Hagen
	Hugo Rader, Solingen
	Ernst Hugo Rasspe, Solingen
	Rau & Koch , Südwest
	Erhardt Reich, Schweina bei Liebenstein
	"Reimers" P.J.V. Antrieb Werner Reimers KG, Bad Homburg
	"REMO" Gebrüder Rempt, Suhl
	Cuno Remscheid & Co., Auf der Höhe

	Josef Reuleaux, Solingen-Wald
	Rheinisch-Westfälische Sprengstoff Aktiengesellschaft, Werk Stadeln, bei Fürth, (acquired later by Alfred Nobel & Co, then in turn even later by Dynamit Nobel AG)
	Rheinisch-Westfälische Sprengstoff Aktiengesellschaft, Werk Stadeln, bei Fürth, (acquired later by Alfred Nobel & Co, then in turn even later by Dynamit Nobel AG) [variation]
	Rheinmetall AG, Sömmerda [early]
	Rheinmetall AG, Sömmerda
	Rheinmetall-Borsig AG, Sömmerda [variant, transition]
	Rheinmetall-Borsig AG, Sömmerda [later]
	Rheinmetall-Borsig AG, Sömmerda [variant]
	Rheinmetall AG, Düsseldorf [post-WWII]
	Kuno Ritter, Solingen
	Gebr. Röchling, Ludwigshafen a. Rhein [early]
	Röchling´sche Eisen- u. Stahlwerke GmbH, Wetzlar [late]

	G. Rodenstock Optische Werke, München
	Röhm-Gesellschaft, Präzisionswerkzeug-Fabrik, Zella-Mehlis
	Röhm GmbH, Sontheim [post-WWII]
	Röhm GmbH, Sontheim [post-WWII]
	Ernst Römer, Solingen
	Römerwerk AG, Suhl
	Romüso, Merscheid
	August Rother, Solingen
	Ernst Röttgen, Solingen
	Rottweil Pulverfabrik, Rottweil
	Rottweil Pulverfabrik, Rottweil [post-WWII]
	Rottweil Pulverfabrik, Rottweil [post-WWII]

RUKA Rathenow	"RUKA" Runge & Kaulfuss, Rathenow
S	"S" Simson & Co., Suhl [variant]
S	"S" Simson & Co., Suhl [variant]
S	"S" Simson & Co., Suhl [variant]
	J.P. Sauer & Sohn Gewehrfabrik, Suhl [early]
Sauer.	J.P. Sauer & Sohn Gewehrfabrik, Suhl [early]
S&S	J.P. Sauer & Sohn Gewehrfabrik, Suhl [early]
S&S	J.P. Sauer & Sohn Gewehrfabrik, Suhl [early]
	J.P. Sauer & Sohn Gewehrfabrik, Suhl
SAUER	J.P. Sauer & Sohn Gewehrfabrik, Suhl [variant]
Original Sauer	J.P. Sauer & Sohn Gewehrfabrik, Suhl [post-WWII]
ORIGINAL SAUER	J.P. Sauer & Sohn Gewehrfabrik, Suhl [post-WWII]

	J.P. Sauer & Sohn Gewehrfabrik, Suhl [post-WWII]
	"SB" Sellier & Bellot Patronen-, Zündhütchen- u. Metallwarenfabrik AG, Prag
S C	"SC" R. & G. Schmöle & Co., Metallwerke, Menden
	E.D. Schaaff, Solingen
EUGEN SCHEIDT SOLINGEN-OHLIGS	Eugen Scheidt, Solingen-Ohligs
SCHEINTOD REPETIER PISTOLE	"SCHEINTOD" Burgsmüller & Sohn, Kreiensen/Harz
	Carl Schlieper, Solingen
JOS. SCHLIMBACH SOLINGEN	Josef Schlimbach, Solingen
	Hugo Schmeisser/C.G. Haenel, Suhl
	Schmidt & Bender, Biebertal/Wetzlar [post-WWII]
CARL SCHMIDT SOHN A-G. SOLINGEN	Carl Schmidt Sohn AG, Solingen Herbert Schmidt Waffenfabrik, Zella-Mehlis
	J. A. Schmidt & Sohn, Solingen

	J. A. Schmidt & Sohn, Solingen [variant]
	J. A. Schmidt & Söhne, Juliuswerk, Solingen
	Rudolf Schmidt, Solingen
	Friedrich August Schmitz, Solingen
	Hermann Schneider, Solingen- Aufderhöhe
	Schneider & Co. KG, Optische Werke, Werk Göttingen and Werk Bad Kreuznach
	Schnittert Abr, WASSO, Solingen-Wald
	Emil Schrick & Sohn, Solingen
	August Schüler, Suhl
	Schütz, Ruf & Co, Kassel
	A.W. Schwarzlose, Berlin
	A.W. Schwarzlose, Berlin [variant]

	W&H Seibert, Wetzlar
	Paul Seilheimer, Solingen
	Paul Seilheimer, Solingen [variant]
	Paul Seilheimer, Solingen [variant]
	"Senta" Hermann Linder & Söhne, Solingen
	Hugo Servatius, Solingen
	Siemens & Halske A.G., Berlin
	Friedrich Siemens-Werke AG, Unternehmung für Wärmetechnik und Gasapparatebau, Wien
	"SIG" Schweizerische Industrie Gesellschaft, Neuhausen, Switzerland
	Otto Simon, Steinbach
	"SIMPLEX" Theodor Bergmann Maschinen u. Waffenfabrik, Gaggenau
	Simson & Co., Suhl

SINOXID	"SINOXID" Rheinisch-Westfälische Sprengstoff Aktiengesellschaft, Werk Stadeln, bei Fürth, (acquired by Alfred Nobel & Co, then in turn even later by Dynamit Nobel AG)
ŠKODA	Škoda Aktiengesellschaft (eh. Skoda-Werke, or Czech *Škodovy závody*), Pilsen
SLAVIA	"SLAVIA", Antonin Vilímec, Kdyne and then later used by Kohout & Spol.(SPK), Kdyne, Czechoslovakia
Slavia	"Slavia" Posumavska Zbrojovka Kohout & Spol. (SPK), Kdyne, Czech. [post-WWII]
SMF	"SMF", Solinger Metallwaffenfabrik, Stoecker & Co., Solingen
SOLINGER AXT- UND HAUERFABRIK	Solinger Axt und Hauerfabrik, Solingen
SOMET	"SOMET" Meopta, Prerov, Czechoslovakia [post-WWII]
SPALTENEDER · MÜNCHEN ·	Spalteneder, München
SPANDAU	Königlich-Preussische Gewehrfabrik Spandau, Berlin-Spandau
	Gustav Spitzer, Solingen
K	Spreewerk GmbH, Metallwarenfabrik, Berlin-Spandau[59]
	"Springer" Carl Wusthof (Gladiatorwerk) AG, Solingen

SRBASTYS PRAHA	Srb and Stys, Fabrik Praziser Messinstrumente, Prague, Czechoslovakia
SRBASTYS PRAHA	Srb and Stys, Fabrik Praziser Messinstrumente, Prague, Czechoslovakia [variant]
STAR TRADE MARK	"STAR" Star Bonifacio Echeverria SA, Eibar, Spain
STAR TRADE MARK	"STAR" Star Bonifacio Echeverria SA, Eibar, Spain [variant]
STAR ARMAS FIREARMS EIBAR ESPAÑA	"STAR" Star Bonifacio Echeverria SA, Eibar, Spain [post-WWII]
STEINHEIL OPTIK	"Steinheil" Optische Werke C. A. Steinheil Söhne GmbH., München
FRANZ STEINHOFF MOBIL SOLINGEN-WALD	Franz Steinhoff, Solingen-Wald
ST.W	Stenda Werke Waffenfabrik, Suhl
STENDA	Stenda Werke Waffenfabrik, Suhl [variant]
STEYR	Steyr-Daimler-Puch AG, Steyr
	Steyr-Daimler-Puch AG, Warschau [suspected]
S	Steyr-Solothurn Waffen AG, Steyr

	Franz Stock, Maschinen- u. Werkzeugfabrik, Berlin-Neukölln
	Franz Stock, Maschinen- u. Werkzeugfabrik, Berlin-Neukölln [variant]
	"Stocko" Metallwarenfabriken OHG, Wuppertal-Elberfeld
	Stoewer-Werke AG, Kraftfahrzeugfabrik, Stettin
	Otto Stöver, Solingen
	Fr. Stübgen & Co., Metallwerk, Erfurt
	Süddeutsche Messerfabrik, Gefrees
	"Suhl Consortium" consisted of C.G. Haenel, J.P. Sauer & Sohn, and V. Chr. Schilling [see text]
	D. Swarovski, Glasfabrik u. Tyrolit-Schleifmittelwerke, Tyrol, Austria
	"THÜRINGEN" Waffengenossenschaft Thüringen
	Karl Tiegel, Riemberg-Breslau
	"TIGER" Lauterjung & Sohn GmbH & Co. Tigerwerk, Solingen

	"TIGER" Lauterjung & Sohn GmbH & Co. Tigerwerk, Solingen
	Carl Tillmans & Söhne, Solingen
	Tiroler Waffenfabrik Johann Peterlongo, Innsbruck, Austria
	Tiroler Waffenfabrik Johann Peterlongo, Innsbruck, Austria [variant]
	Gebr. Torley, Solingen
	"Torpedo" Fichtel & Sachs, Schweinfurt
	Torpedo-Werke AG, Fahrräder u. Schreibmaschinenfabrik, Frankfurt/Main-Rödelheim
	"TW" VEB Fahrzeug- und Jagdwaffenwerk Ernst Thälmann, Suhl, German Democratic Republic [post-WWII]
	"TW" VEB Fahrzeug- und Jagdwaffenwerk Ernst Thälmann, Suhl, German Democratic Republic [post-WWII]
	"U" Rheinisch-Westfälische Sprengstoff Aktiengesellschaft, Karlsruhe
	"Undine" Kuno Meisenburg, Solingen
	"UNIQUE" Manufacture d'Armes des Pyrenees Francaises (MAP), Hendaye, France

Logo	Manufacturer
	"UNIQUE" Manufacture d'Armes des Pyrenees Francaises (MAP), Hendaye, France [variant]
7.65 m/m 9 SCHUSS	"UNIQUE" Manufacture d'Armes des Pyrenees Francaises (MAP), Hendaye, France [1940-1944 variant]
	"V" Vollmer Weke GmbH, Biberach an der Riß
VAF / C	"VAF" Vereinigten Automaten Fabriken Pelzer und Cie, Cöln-Ehrenfeld
VDM	"VDM" Vereinigte Deutsche Metallwerke AG, Zweigniederlassung der Süddeutschen Metallindustrie, Frankfurt & Nürnberg
Venditor	"Venditor" Dynamit-Actien-Gesellschaft vormals Alfred Nobel & Co., Abteilung Celluloid-und-Kunststoff-Fabrik, Werk Troisdorf (Bez. Köln); verkauf durch Venditor, Kunststoff-Verkaufsges. m.b.H, Troisdorf (Bez. Köln)
	"VENUS" Oskar Will, Venus-Waffenwerk; Zella-Mehlis, Thüringen
	Vereinigte Oberschlesische Hüttenwerke AG, Werk Andreashütte, Andreashütte
	"VIS" Fabryka Broni W. Radomiu, Radom, Poland
	Eduard Vitting, Solingen
	Voigtländer & Sohn AG, Optisches Gerät, Werk Braunschweig [early]
	Voigtländer & Sohn AG, Optisches Gerät, Werk Braunschweig

	Adolf Völker, Zangenfabrik, Schmalkalden
	Adolf Völker, Zangenfabrik, Schmalkalden [variant]
	F. von Brosy Steinberg, Solingen
	Emil Voos Waffenfabrik, Solingen
	Gustav Voss, Solingen
	"W" Thüringische Maschinen- und Fahrradfabrik Walter & Co. GmbH, Mühlhausen
	"W" Wichmann, Gebr., Zeichengeräte, Vermessungsinstrumente, Berlin
	"WAFFENHAMMER" Max Kaiser Waffenhammer, Deggendorf
	Waffen Jung, Stuttgart
	Waffenstadt Suhl, Suhl
	I.C. Wagner & Co. GmbH, Heyerode
	Wilhelm Wagner, Solingen-Merscheid

Trademark	Manufacturer
SternPistole A.W.Z.	Albin Wahl, Zella-Mehlis
OTTO WALTER Nº 11550	Otto Walter, Fürstenfeldbruck
	Carl Walther Waffenfabrik, Zella-Mehlis [early]
DIE WALTHERSCHLEIFE	Carl Walther Waffenfabrik, Zella-Mehlis
WALTHER	Carl Walther Waffenfabrik, Zella-Mehlis
	Fritz Weber & Co., Laternenfabrik, Berlin
THOMAS WEIPUTZ GEGR 1880 SOLINGEN-HÖHSCHEID	Thomas Weiputz, Solingen-Höhscheid
WILH. WELTERSBACH SOLINGEN	Wilhelm Weltersbach, Solingen
	Fritz Werner AG, Maschinen- u. Werkzeugfabrik, Berlin
FRITZ WERNER	Fritz Werner AG, Maschinen- u. Werkzeugfabrik, Berlin [post-WWII]
gegr. 1828 AWERTH SOLINGEN	A. Werth, Solingen
A.WERTH SOLINGEN	A. Werth, Solingen [variant]

	Wester & Butz, Solingen
	Feinwerkbau Westinger & Altenburger, Oberndorf [post-WWII]
	Gebrüder Weyersberg, Solingen
	Gebrüder Weyersberg, Solingen [variant]
	Gottfried Weyersberg, Solingen
	Gustav Weyersberg Nachfolger, Solingen
	"WKC" Weyersberg, Kirschbaum und Cie., Waffenfabrik GmbH, Solingen [early]
	"WKC" Weyersberg, Kirschbaum und Cie., Waffenfabrik GmbH, Solingen
	"WKC" Weyersberg, Kirschbaum und Cie., Waffenfabrik GmbH, Solingen
	"WMW" Waffenfabrik Max Weyersberg, Solingen
	"WMW" Waffenfabrik Max Weyersberg, Solingen [variant]
	Max Weyersberg, Solingen [variant]

	Paul Weyersburg & Co. Waffenfabrik, Solingen
	Reinhard Weyersberg, Solingen
	Wiener Waffenfabrik, Wien
	Anton Wingen Jr., Solingen
	R. Winkel GmbH Optische Industrie, Göttingen
	Gustav Wirth, Solingen-Grafrath
WISSNER	Metallwarenfabrik AG (eh. H. Wissner), Zella-Mehlis
	Metallwarenfabrik AG (eh. H. Wissner), Bensheim [post-WWII]
	Ernst Erich Witte, Solingen
	Ernst Erich Witte, Solingen [variant]
	P.D. Witte, Solingen
	Dr. F. A. Wöhler, Kassel

JOSEF WOLF MÜNCHEN-AU	Joseph Wolf Waffenfabrik, München
IS. WOLFERTZ SOLINGEN	Isidor Wolfertz, Solingen
WURAG	"WURAG" Eisen- u. Stahlwerke AG, Hohenlimburg
	"WURAG" Eisen- u. Stahlwerke AG, Hohenlimburg [variant]
ED. WÜSTHOF SOLINGEN	Eduardt Wüsthof Drezackwerk, Solingen
Z	"Z" Clemen und Jung, Solingen
Z	"Z" Praga Zbrojovka, Prague and later Ceskoslovenska Zbrojovka, Brno
CARL ZANDER KWICK SOLINGEN	Carl Zander, Solingen
	Ludwig Zeilter
Zeiss	Carl Zeiss, Optische Geräte / Militärabteilung, Jena [early]
CARL ZEISS JENA	Carl Zeiss, Optische Geräte / Militärabteilung, Jena
CARL ZEISS	Carl Zeiss, Jena, German Democratic Republic [post-WWII]

Zico	Zimmermann & Co., "Zico", Preß- u. Stanzwerk, Remscheid
Zi - Di	"Zi-Di" Heinrich Diem Gewehrfabrik, Benshausen

Appendix II: Markings Used Internally by Various Manufacturers

Various manufacturing, acceptance and internal control markings can be confused with trademarks and logos. Manufacturers often used one or more numbers, letters, proof or inspection marks within their organizations to denote a stage of completion or acceptance. Some denoted a subsidiary origin. These should not be confused for trademarks or logos per se, but had specific meanings within the company. Those observed are presented here for clarity. Some of the most commonly encountered are shown below. As there are countless different markings one might encounter on WWI-WWII optics, edgeware and firearms, this list should by no means be considered complete.

Fig. 186: A barrel manufactured from a high-quality steel blank supplied by Sächsische Gußstahlwerke Döhlen AG, Freital and bearing a company mark that probably indicates a lot or series. This barrel was likely finish-machined by Alkett (Altmärkisches Kettenwerk GmbH, later known as "Maget"), Berlin-Tegel as indicated by the additional "cra" code. Steel and barrel makers used a number of symbols and marks that were peculiar to a specific firm, but can really not be classified as true trademarks or logos. *Courtesy Kenneth Huddle.*

	[asterisk] Generally believed to represent a part or component that did not initially meet specifications; but was salvageable with additional work
	[plus sign] Generally used to represent an extra or spare in regard to firearms, such as a spare magazine.
	[flaming bomb] Deutsche Waffen- und Munitionsfabriken AG, Berlin [final factory inspection mark]
	Steyr-Daimler-Puch AG, Steyr [internal department of acceptance mark]
	Carl Walther Waffenfabrik, Zella-Mehlis [internal department or acceptance mark]
	Carl Walther Waffenfabrik, Zella-Mehlis [internal department or acceptance mark]
	Carl Walther Waffenfabrik, Zella-Mehlis [internal department or acceptance mark]
	Carl Walther Waffenfabrik, Zella-Mehlis [internal department or acceptance mark]
	Carl Walther Waffenfabrik, Zella-Mehlis [internal department or acceptance mark]
	Carl Walther Waffenfabrik, Zella-Mehlis [internal department or acceptance mark]
	Carl Walther Waffenfabrik, Zella-Mehlis [internal department or acceptance mark]
	"bnz" Steyr-Daimler-Puch AG, Steyr
	"byf" Mauser-Werke AG, Oberndorf a.N. [later variant]
	"C" acceptance mark used by Česká Zbrojovka, Brno, Czech.

CrV	Fritz Werner AG, Maschinen- u. Werkzeugfabrik, Berlin
dot 11	"dot" Waffenwerke Brünn I, Werk Brünn (Česká Zbrojovka, Brno, Czech.)
dou 1	"duv" Berlin-Lübecker Maschinenfabrik, Lübeck
	Carl Eickhorn, Solingen [small arms components]
EE	"ERMA" Erfurter Maschinenfabrik Berthold Geipel GmbH, Erfurt
H	C.G. Haenel Waffen- und Fahrradfabrik AG, Suhl
H	C.G. Haenel Waffen- und Fahrradfabrik AG, Suhl
H	C.G. Haenel Waffen- und Fahrradfabrik AG, Suhl
H	Hirtenberger Patronen- und Zündhütchenfabrik, Hirtenberg
K	Spreewerk GmbH, Metallwarenfabrik, Kratzau, Czechoslovakia
M	Mauser-Werke, Oberndorf a/Neckar
MI	"MR" internal control mark used internally by Fabrique Nationale D'Armes de Guerre, Herstal Belgium, under German WWII occupation [variations exist]
MR	"MR" internal control mark used internally by Fabrique Nationale D'Armes de Guerre, Herstal Belgium, under German WWII occupation
N	"N" used to mark finished components made by Berlin-Lübecker Maschinenfabrik, Lübeck or one its sub-contractors

	"qve" Berlin-Lübecker Maschinenfabrik, Lübeck
	Simson & Co., Suhl [may represent a firearm over-hauled by Simson]
	Sächsische Gußstahlwerke Döhlen AG, Freital
	"V" acceptance mark used by Gustloff-Werke or one of its sub-contactors
	Steyr-Daimler-Puch AG, Warschau [suspected to represent Steyr, Warschau]
	"WR" Mauser-Werke, Oberndorf a/Neckar internal mark. Abbreviation for "Werks-Revision", indicating a component that required re-inspection. Also used as an acceptance mark on weapons produced under French occupation 1945-1946
	"Z" Praga Zbrojovka, Prague and later Ceskoslovenska Zbrojovka, Brno

Appendix III: Miscellaneous Markings Used by Various Official Government Organizations

Various organizational, official and/or governmental markings can be confused with traditional trademarks and logos. Agencies used a multitude of markings to mark property and/or denote origin, proof firing, approval, acceptance or possession. Some of the most commonly encountered are shown below. As there are countless different markings one might encounter on WWI-WWII optics, edged weapons and firearms, this list of commonly encountered marks should by no means be considered complete and is only included for clarity and quick reference.

Fig. 187: Commercial, non-military markings and proof on a prototype K98v made by Mauser near the end of WWII. The numerals indicate the date of production (March 1945), the antler represents the Württemberg proof house (used by the Ulm proof house after WWII) and the National Socialist eagle over the letter "N" is the commercial smokeless powder proof. *Courtesy Jon Speed.*

	French proof denoting a finished and accepted firearm. Used after WWII by Mauser-Werke which made small arms under French occupation, 1945-1946
	"MPD" stylized abbreviation for the "Staatliches Materialprüfungsamt" or "State Materials Testing Office" [1]
	Police inspection/acceptance/property mark
	German optical lubricant marking; also used to denote a spare component or assembly [WWII]
	German optical lubricant marking; also used as an index marking [WWII]
	German optical lubricant marking [WWII]
	Manufacturing mark used in the DDR (German Democratic Republic) [post-WWII]
	Soviet arsenal mark [in the context of this text the marking is frequently found on German WWII firearms captured by the Soviets, overhauled in USSR arsenals and later exported to the West for commercial sale post-WWII- it frequently appears as an "X"]
	Berlin-Treptow Proof House commercial proof [pre-1945]
	Suhl Proof House commercial proof [used pre-1945 and with the addition of an "M", "N" or "SP" by the DDR after WWII]
	Württemberg Proof House commercial proof [this mark was also used after WWII by the Ulm Proof House]
	Zella-Mehlis, Thüringen Proof House commercial proof [pre-1945]
	Zella-Mehlis, Thüringen [thought to be a location identifier for the proof house or perhaps a commercial smoothbore proof from the same facility]
	German shotgun/smoothbore gauge marking, seen on shotguns, flare guns, etc. Other numerals also observed such as "12", "14", "16" and "20"

4 UT	Italian proof used at the Pietro Beretta facility in Gardone which fell under the under Republic of Socialist Italy/German occupation; indicates German acceptance/ quality control
B 74	Heereswaffenamt transitional inspection mark, used up to ca. 1935 at some small arms facilities [numerous variations of letters and numbers]
	Imperial German military inspector or "Beschußmeister" mark; numerous variations and possible Fraktur letters, e.g. "A", "B", "C", "D", "FJ", "P", "X" etc.
D. R. 2422	"DR" abbreviation for "Deutsches Reich"
DRP	"DRP" abbreviation for the German Postal Service, or "Deutsches Reichs Post"
EWB	"EWB" stylized abbreviation for the " Einwohnerwehr Bayern", a right-wing, anti-communist militia disbanded in 1921
	Austrian firing proof mark, or "Beschusstempel"
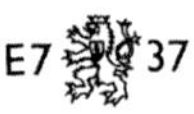	Pre-WWII Czechoslovak proof; numerals indicate location and date of acceptance
	Post-WWII Czechoslovak proof; also used as property mark pre- and post-WWII with slight differences
	Imperial German military firing proof mark, or "Beschusstempel" [DWM version]
	Imperial German black powder barrel firing proof mark, or "Beschusstempel" [1891-1912 variant]
	Post-WWI "Weimar" German military Ausrustungsamt inspection proof
	Weimar/Early National Socialist military firing proof mark, or "Beschusstempel"
	National Socialist military firing proof mark, or "Beschusstempel"

	National Socialist military firing proof mark, or "Beschusstempel" [variant]
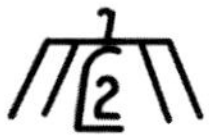	"Luftamt" Luftwaffe military inspection mark [numerous variants]
	National Socialist era Police inspection mark [numerous variants]
	Imperial German Naval inspection and property mark
	Weimar Naval property mark Socialist Kriegsmarine property mark
	National Socialist Kriegsmarine property mark [numerous variants]
	Post-1912 German smokeless powder commercial proof
	National Socialist smokeless powder commercial proof [post-1940]
	DDR smokeless powder proof [post-WWII]
	BRD (Federal Republic of Germany) smokeless powder proof [post-WWII]
Nitro	"Nitro" a commercial marking indicating that a firearm meant for export was made to use modern, smokeless "Nitro-Cellulose" gun powder; usually used in conjunction with the Crown/N or Eagle/N markings noted above.
Nitro	"Nitro" [script variant] a commercial marking indicating that a firearm meant for export was made to use modern, smokeless "Nitro-Cellulose" gun powder; usually used in conjunction with the Crown/N or Eagle/N markings noted above.
	Austrian smokeless powder proof, Ferlach
	Czech smokeless powder proof, Prague [ca.1899-1931]

NPV	Austrian smokeless powder proof, Wien [Vienna]
NPW	Czech smokeless powder proof, Weipert [ca.1899-1931]
NSKK	"NSKK" Nationalsozialistisches Kraftfahrkorps", or National Socialist Transportation Corps, property mark
PDM	"PDM" Police Directorate Munich, or "Polizei Direktion München", property mark
PTV	"PTV" Police Traffic and Technical School, or " Polizeischule für Technik und Verkehr " re-work inspection mark
RAJD	"RAD" Reichs Arbeits Dienst, or "Empire Labor Service", marking used on RAD edged weapons and implements
RFV	"RFV" abbreviation for the German Finance Ministry, or "Reichs-Finanzverwaltung"
R.J.	"RJ" abbreviation for the German Justice Ministry, or "Reichs-Justizministerium"
RZM	"RZM" stylized abbreviation for "Reichszeugmeisterei", or very loosely, "National Material Control", applied to items procured by the "Party" (NSDAP) organizations during the National Socialist era
SA	"SA" Finnish military property mark [Suomen Armeija]
	"SA" stylized abbreviation for the NSDAP Sturmabteilung, or "Storm Troops"- frequently seen stamped or engraved on small arms and equipment issued by that organization.
	"S" stylized-single rune frequently seen stamped on small arms and equipment belonging to the Hitler Youth or the Waffen-SS unit "Hitlerjugend"
	"SS" stylized abbreviation for the NSDAP Schutzstaffel, or "*SS*"; frequently seen stamped or engraved on small arms and equipment issued by that organization

	"SS" stylized abbreviation for the NSDAP Schutzstaffel, or "*SS*"; frequently seen stamped or engraved on small arms and equipment issued by that organization
	"VDE" Not a manufacturer's trademark but a representation of the Verband Deutscher Elektrotechniker (German Federation of Electrical Engineers)
	"WaA" German military inspection mark, or "Wehrmachtabnahmestempel". Usually followed by numerals or a letter/numeral combination [numerous variants]
	"WaA" German military inspection mark, or " Wehrmachtabnahmestempel" [numerous variants]

Notes

1 What is also missing at the time of this writing is a comprehensive and accurate reference documenting and explaining the German wartime Reichs Betrieb numbers, or "RB" numbers.

2 After the US Civil War, Texas rancher Samuel Augustus Maverick decided that since all other cattle he knew of were branded, his would be identified by having no mark or brand at all. Thus, the term "maverick", originally referred to an unbranded calf.

3 The only major exception is the Württemberg region, and specifically the area in and around Oberndorf which became known for the expansive Mauser-Werke AG.

4 Other towns with government arsenals in their midst included Amburg in Bavaria, Danzig, and Berlin-Spandau.

5 CIOS XXIX-52, *German Optical Production*, pg. 3.

6 CIOS XXIX-52, *German Optical Production*, pg. 3.

7 This should not be confused with national commercial or military proof markings; usually consisting of a crown or eagle over a letter or the National Socialist "swastika".

8 See Weaver, W. Darrin, et al: *Mauser Pistolen*, pg. 36.

9 For instance, Walther was assigned the military letter code "ac" which it used on most all of its firearms destined for the military, but was also assigned RB Nr. 0/0345/0064. JP Sauer was assigned "ce" and 0/0345/0044, Berlin-Lübecker Maschinenfabrik "duv", "qve" and 0/0499/0015, Mauser "S42", "42", "byf", "svw" and 0/0762/0062 and lastly Gustloff "bcd" and 0/0342/0001.

10 See Huddle, Kenneth: *An Overview of German WWII Markings*, p. 8.

11 German cutlery manufacturers proudly placed their logos and maker marks on the reverse of their blades up to about 1935-36. After that, manufacturers reluctantly began to replace their names and logos with the RZM codes assigned to them by the *Reichszeugmeisterei*, on blades destined for political organizations such as the SA, SS, Polizei, etc. Blades destined for the military were marked with the traditional "S" codes, number codes and then all were eventually supposed to be marked with the military letter code system. Compliance with this was spotty, as evidenced by numbers of 98K bayonets with traditional maker's names and marks well into WWII. Dagger and sword production then essentially ceased in 1941-1942. Commercial cutlery continued to be marked in the pre-war manner.

12 Firearm prototypes and presentation weapons were frequently and proudly marked with the logo or trademark of the development firm, even during the National Socialist era; this, a method of garnering favor for the company with acceptance officials.

13 See CIOS XXVI-26, *Voigtländer und Sohn A.G. Braunschweig-Gliesmarode*, pg. 4.

14 See Shomper, Paul: *Bavarian Arsenal Amberg*, pgs. 19-23.

15 Weaver, W. Darrin: Hitler's Garands, pg. 215.

16 CIOS XXX-14, *Visit to Thuringia Small Arms Industry*, pg. 4.

17 Many collectors refer to rifles made in/by Ceskoslovenska Zbrojovka as "BRNO" rifles- a bit of a misnomer. After German occupation, the Brno facility produced arms marked with the "dot" military letter code for the Wehrmacht.

18 Waffenwerke Brünn fell under Škoda of the "Reichswerke Herman Göring AG" and by 1941 consisted of the following firms: Waffenwerke Brünn AG, Brünn, Optikotechna GmbH, Prerau, Patronenhülsen- u. Metallwarenfabrik AG, Rokytzan, AG vorm. Skodawerke, Pilsen (a little over half-controlled), Eisenwerke Podbrezove, Explosia AG, Farola rum. Fabr. f. gezog. Stahl AG, Bukarest, Magneton GmbH, Kremsier, Detona GmbH, Prag, Metrom 1. rum. Metallw. Fabrik, Bukarest, Montania Berg- u. Hütten AG, Preßburg, Radioslavia AG, Prag, Rudnica A.D., Belgrad, Sandrik Silber- u. Metallwaren AG, Prag, Sandrik Silber- u. Metallwaren AG, Dolnie Hamry, Stadion Fahrradfabrik AG, Prag, Synthesia Chem. Werke AG, Prag, and UMA Kunststofferzeugung AG, Prag.

19 After World War II all the Czech firearms manufacturers were managed by a single central marketing agency (Omnipol) which required that any firearm meant or sold for export would bear the BRNO marking.

20 See Shomper, Paul: *Prussian Arsenal Danzig*, p. 25-27.

21 See Shomper, Paul: letter to the author dated November 16, 2009.

22 In 1933, DIWAG was itself absorbed into the state-controlled VIAG (Vereinigte Industriewerk AG. See Balcar, Jan and Ron Clarin: *The P.38 Pistol Spreewerk Production*, pg. 8.

23 CIOS XXIX-52, *German Optical Production*, pg. 3.

24 Of note, Baron Paul von Gontard, the manager of DWM at the time of the Quandt takeover married Clara Busch in 1897. Clara was the daughter of the German-American brewer, Adolphus Busch. The Quandt Group asked him to resign his position at DWM, which he did. Gontard subsequently joined his wife in the United States and became involved in the Anheuser-Busch (Budweiser brand) company.

25 DWM Posen also made a small number of VG1Volksgewehr rifles near the end of the war.

26 See *Dynamit A.G., Troisdorf,* CIOS XXXI-3, pg. 3.

27 See *Dynamit A.G.,* CIOS XXI-3, pgs. 3-5.

28 See Shomper, Paul: letter to the author dated November 16, 2009.

29 The Fortuna tradename was also used to market motorcycles
in the 1920s by the "Fortuna-Kleinmotorenfahrzeugwerk Nürnberg".

30 See *Spezial -Katalog Nr. 44 Moderne Waffen* (1929). Of note, after WWII, Geco and all of its assets were eventually acquired by the RUAG holding company, which also eventually acquired RWS, Rottweil and other ammunition makers.

31 Swiss Gauleiter Wilhelm Gustloff was shot dead by a Jewish student, David Franfurter, on February 4, 1936. The Nazi regime promptly made Gustloff into a martyr, and he was eulogized by Hitler, himself. See Weaver, W. Darrin: *Hitler's Garands*, pg. 100.

32 See *Carl Zeiss AG*, http://en.wikipedia.org/wiki/Hensoldt

33 Original members included Bayer, BASF, Agfa, Griesheim-Elektron, Weilerter Meer, and Hoechst.

34 Shomper, Paul: letter to the author dated November 16, 2009.

35 Krieghoff's Kufstein facility was reorganized after WWII as the "Tiroler Maschinenbau u. Holzindustrie, Kufstein, Tirol".

36 See CIOS XXXIII-54 *Organization of Ernst Leitz at Wetzlar*, pgs. 1-23.

37 See Weaver, W. Darrin, Speed, Jon, et al: *Mauser Pistolen*, pg. 35.

38 The actual system was originally developed for the Greek Army in 1903.

39 See Weaver, W. Darrin, Speed, Jon, et al: *Mauser Pistolen*, pg. 36.

40 See Weaver, W. Darrin: *The Mauser Barrel*. Mauser's trademark is often described as a "banner" by collectors, but this is not at all correct. Mauser's trademark was patterned after its DWM parent company's trademark, a stylized "Tonne" powder barrel.

41 See Myrvang, Folke: *MG34-MG42 German Universal Machineguns*, pg. 21.

42 Opticotechna was nominally supervised by the Austrian firm Feinapparate-Bau GmbH during WWII.

43 See Rohan, Stephen, *Eyes of the Wehrmacht*, Vol. I, pg 14.

44 See Shomper, Paul: *Reichswerke "Hermann Göring" AG.*

45 Rheinmetall-Borsig fell under the "Reichswerke Herman Göring AG" and by 1941 consisted of the following firms: Aceros y Metales S.A., Barcelona, Alkett GmbH, Berlin, Busch-Jaeger-Lüdenscheid, Eisen- und Metall AG, Essen, Gemeinn. Bauverein Volkswohl, Sömmerda, Gemeinn. Wohnungs- u. Siedlungsgesellschaft, Berlin.-Tegel, Graftenberger Werkshandelsges. GmbH, Berlin, Hydraulik GmbH, Duisburg, Patronen, Zündhütchen und Metallwarenfabrik AG, Schönebeck, Rheinmetall-Borsig Cia, Buenos Aires, Rheinmetall-Borsig Handelsges. mbH, Berlin, Rheinmetall-Borsig S.A., Rome, Rheinmetall-Solo Vertriebs GmbH, Berlin, Rheinwohnungsbau AG, Düsseldorf, Agrara AG, Bukarest, Bankhaus Simons & Co. KG. Düsseldorf, N.V. Comprimo, Amsterdam, Düsseldorf Spar- und Bauverein, Düsseldorf Gemeinschaft Bremerhaven, Industriefinanzierungs AG Ost, Berlin, Schiffsverkaufges. GmbH, Berlin Thüringische Zellwolle AG, Schwarza, Tornescher Holzzuckerverwertg. und Studien GmbH, Tornesch, and Weksspoor N.V., Amsterdam .

46 See ETO Technical Intelligence Report Number 406, *Volkssturm Experimental Semi Automatic Carbine Cal. 7,92 mm Short.*

47 Škoda Aktiengesellschaft (eh. Skoda-Werke, or Czech *Škodovy závody*), Pilsen fell under the „Reichswerke Herman Göring AG" and by 1941 consisted of the following firms: AG. vorm. Škodawerke, Pilsen; ABAG, Afghan. Bauges., Kabul, AVIA AG, Prag, Böhmische Kontaktwerke AG, Komotau, Böhm.-Mähr. Schrottvereinigung, Prag, Götzl & Schmidt GmbH, Prag, Škoda AD, Belgrad, KABLO AG, Kabelfabrik, Prag, Konstruktiva Bau AH Steyr- Škoda, Bulg. Autohandels-AG, Sofia, Novak & Jahn, Brückenbau, Prag, OMNIPOL AG, Prag, Wien, and London, Oy Škoda VB, Helsingfors, Polnische Ges. d. Škodawerke GmbH, Warschau, Sarex, SA, Bukarest, Škoda Brasieiera SA, Rio de Janeiro, Škoda India Ltd., Bombay, Škoda Industriale Romana, Bukarest, Škoda Malaya Ltd., Singapore, Škoda works of South Africa Ltd., Johannesburg, SA Škoda Platense, Buenos Aires, Soc. A. Ind. Italiana, Rome, Soc. Ir. Škoda, Teheran, Tschechische Luftverkehrsgesellschaft AG, Prag, Unizele de Fier, Bukarest, Unizele Metalurgice, Bukarest, Unizele Metalurgice Ploesti, Bukarest, and Waffenwerke Brünn AG, Brünn.

48 See Mathisen, Per: Email correspondence to the author dated October 18, 2001.

49 See Blazicek, Andrej: Email correspondence to the author dated April 20, 2004.

50 Another source states that production was shifted to "rented buildings" near Spreewerke, Grottau. See Balcar, Jan: Email correspondence to the author dated May 15, 2004.

51 The firm's name is spelled in the singular "Spreewerk" (or Spree Work) in contemporary company literature and documents and less frequently spelled in government documents in the plural, "Spreewerke" (Spree Works).

52 Steyr fell under the "Reichswerke Herman Göring AG" and by 1941 consisted of the following firms: Steyr-Daimler-Puch AG, Wien, Steyr-Skoda-Bulg. Autohandels AG, Sofia, Gemeinn. Wohnungsges. Steyr-Daimler-Puch GmbH, Buges. für Arbeiter-Wohnh'user Wiener-Neustadt, W-Neustadt, Kromag AG für Werkzeug- und Metallindustrie, Hirtenberg, Nibelungenwerk GmbH, St. Valentin, Romano Germana SAR, Bukarest, Steyr-Austro-Daimler-Puchwerke Ungar. Handels AG, Bukarest, Steyr-Daimler-Puch Handelsges., Zagreb, Steyr-Daimler-Puch Vertr. AG, Krakau, Steyr-Solothurn Waffen-AG, Zürich, Steyr-Daimler-Puch, Warschau and the Vertriebs-AG d. Steyr-Daimler-Puch AG, Zürich

53 See Shomper, Paul: letter to the author dated November 16, 2009, pg 6.

54 See Kersten, Manfred: *Walther A German Legend*, pg 27.

55 CIOS XXX-14, *Visit to Thuringia Small Arms Industry*, pg. 4.

56 History of WKC accessed on-line November 16, 2008, available at http://www.wkc-solingen.de/newshop/history.html

57 Note: the circle within a triangle mark was also used by other manufacturers. The marking was also used by a German watch maker and as an Austro-Hungarian small arms proof/acceptance stamp ca. 1910-1920.

58 The Staatliches Materialprüfungsamt or State Materials Testing Office, located in Dahlem near Berlin (now a district of Berlin itself), abbreviated "MPD", was the responsible authority tasked with monitoring, inspecting and enforcing the German industrial standards. The agency was led by President-Professor Sieber, and Professor Kindsher served as both Vice-President and chief of the "organic" section, which included plastics, rubber, paints, artificial fibers, and glues. The MPD marking is usually found on items made of plastic. Each article was supposed to be marked or have a marking incorporated in the molding process to demonstrate which company made it, exactly what it was made of, and that it was produced to standards. The marking was composed of the stylized combination of the letters "M", "P" and "D" was developed in 1938 . A logo, abbreviation or code representing the manufacturer was to appear in the field above the M, while the field below the M was reserved for the type of plastic and filler materials used in the item's manufacture. See Weaver, W. Darrin: *Kunststoff, A Collector's Guide to WWII Plastics and their Markings.*

59 Note: the circle within a triangle mark was also used by other manufacturers. The marking was also used by a German watch maker and as an Austro-Hungarian small arms proof/acceptance stamp ca. 1910-1920.

Bibliography

Publications:

Arfmann, Peter and Rolf Kallmeyer. *J.P. Sauer & Son*, Suhl, Thuringia: Peter Arfmann Publishing House, 2004.

Balcar, Jan and Ron Clarin: *P.38 Pistol, Spreewerk Production, The*, Lakeville, MN: Ron Clarin, 2009.

Ezell, Edward C.: Handguns of the World, New York, NY: Stackpole Books, 1981.

Gargela, J. and Z. Faktor: *Zeichen auf Handfeuerwaffen*, Praha (Prague), Czechoslovakia: Artria Verlag, 1985.

Genschow, Gustav: *Geco Munitions Preisliste 1939*, Berlin: Gustav Genschow & Co., 1939.

Görtz, Joachim and Don L. Bryans: *German Small Arms Markings*, Marceline, MO, Walsworth Publishing Co., 1997.

Handrich, Hans-Dieter: *Sturmgewehr! From Firepower to Striking Power*, Cobourg, Ontario: Collector Grade Publications, 2004.

Heidler, Michael: *Deutsche Fertigungskennzeichen bis 1945*, Bad Ems: Vogt-Schild Deutschland GmbH, 2007.

Kersten, Manfred: *Walther: A German Legend*, Long Beach, CA, Safari Press, 2001.

Law, Richard: *Sniper Variations of the German K98k Rifle*, Cobourg, Ontario: Collector Grade Publications, 1996.

Mathews, J. Howard: *Firearms Identification, Volume I*, Madison, WI: University of Wisconsin Press, 1962.

Mathews, J. Howard: *Firearms Identification, Volume II*, Madison, WI: University of Wisconsin Press, 1962.

Myrvang, Folke. *MG34-MG42 German Universal Machineguns*, Cobourg, Ontario: Collector Grade Publications, 2002.

Pattarozzi, Mack A. *Luftwaffe Gravity Knife*, Atglen, PA: Schiffer Military History, 2006.

Rohan, Stephan. *Eyes of the Wehrmacht*, Bradbury, CA: Rohan Optical Press, 1996.

Scarlata, Paul S. *Mannlicher Military Rifles*, Lincoln, RI: Andrew Mowbray Publishers, 2004.

Senich, Peter. *The German Assault Rifle 1935-1945,* Boulder, CO: Paladin Press, 1987.

Spielauer, Robert. *German Scopes and Mounts 1914-1945*, Flachberg, Austria: Verlag Spielauer, 2007-2008.

Speed, Jon. *Mauser Archive, The*, Cobourg, Ontario: Collector Grade Publications, 2007.

Speed, Jon. *Mauser Smallbores*, Cobourg, Ontario: Collector Grade Publications, 1998.

Walter, John. *German Military Letter Codes*, East Sussex, England: Small Arms Research Publications, 1996.

Weaver, W. Darrin. *Desperate Measures: The Last-Ditch Weapons of the Nazi Volkssturm,* Cobourg, Ontario: Collector Grade Publications, 2005.

Weaver, W. Darrin. *Hitler's Garands: German Self-Loading Rifles of WWII*, Cobourg, Ontario: Collector Grade Publications, 2001.

Weaver, W. Darrin. *Kunststoffe: A Collector's Guide to German WWII Plastics and their Markings*, Atglen, PA: Schiffer Military History, 2008.

Weaver, W. Darrin, Jon Speed, and Walter Schmid. *Mauser Pistolen*, Cobourg Ontario: Collector Grade Publications, 2008.

Government/Corporate Publications:

AKAH. *Katalog AKAH No. 150*, Albrecht Kind, Berlin-Nürnberg, 1932.

DWM. *Munitions-Katalog,* Karlsruhe i. B.: Deutsche Waffen und Munitionsfabriken, 1904.

DWM. *Price-List,* Karlsruhe (Baden): Berlin-Karlsruher Industrie-Werke AG, 1935.

Ernst Leitz GmbH. *Gewehrkollimator Leico*, Ernst Leitz GmbH, Wetzlar, 1942.

Geco. *Spezial-Katalog Nr. 44, Moderne Waffen*, Gustav Genschow & Co. A.-G., Berlin, 1929.

NSDAP. *Der Vierjahresplan*, Zentralverlag der NSDAP, Franz Eher Nachf. GmbH, Berlin, 1937.

SHAEF. G2, Combined Intelligence Objectives Subcommittee. *Dynamit A.G.*, CIOS XXI-3. APO 413, 1945.

SHAEF. G2, Combined Intelligence Objectives Subcommittee. *Dynamit A.G., Troisdorf,* CIOS XXXI-3. APO 413, 1945.

SHAEF. G2, Combined Intelligence Objectives Subcommittee. *German Optical Production*, CIOS XXIX-52. APO 413, 1945.

SHAEF. G2, Combined Intelligence Objectives Subcommittee. *Investigations into the Organization of Ernst Leitz at Wetzlar*, CIOS XXXIII-54. APO 413, 1945.

SHAEF. G2, Combined Intelligence Objectives Subcommittee. *Visit to Thuringia Small Arms Industry*, CIOS XXX-14. APO 413, 1945.

SHAEF. G2, Combined Intelligence Objectives Subcommittee. *Voigtländer und Sohn A.G. Braunschweig-Gliesmarode*, CIOS XXVI-26. APO 413, 1945.

STEYR-DAIMLER-PUCH. *Jagd- und Sportwaffen 1935*, Steyr-Werke AG, Vienna, 1935.

US Army. Headquarters, Theater Service Forces, Office of the Theater Chief of Ordnance. *Volkssturm Experimental Semi Automatic Carbine Cal. 7,92 mm Short,* ETO Technical Intelligence Report 406. APO 757, 1945.

Zeiss AG. Sales catalog Z 650. Carl Zeiss, Jena, 1938.

Periodicals:

Huddle, Kenneth. "An Overview of German WWII Markings". *Karabiner Collectors Network Newsletter*, October 1993.

Shomper, Paul. "Bavarian Arsenal Amberg". *The Military Rifle Journal*, July 2009, 19-23.

Shomper, Paul. "Prussian Arsenal Danzig". *The Military Rifle Journal*, January 2009, 25-27.

Shomper, Paul. "Prussian Arsenal Spandau". *The Military Rifle Journal*, May 2009, 21-26.

Shomper, Paul. "Reichswerke "Hermann Göring" A.G.". *The Military Rifle Journal*, September 2009, 21-24.

Unknown. "Deutsche Luftwacht (NSFK)". *Luftwelt* Jahr 6, Nr. 10. 1939.

Weaver, W. Darrin. "The Mauser Barrel". *The Military Rifle Journal*, April 2008, 18-21.

On-Line Sources:

Carl Zeiss AG [Internet]. Wikipedia; [accessed November 3, 2008]. Available from http://en.wikipedia.org/wiki/Hensoldt.

Firmen [Internet]. Landesarchiv Nordrhein-Westfalen [accessed December 28, 2009]. Available from http://www.archive.nrw.de/LAV_NRW/jsp/bestand.jsp?archivNr=4&tektId=53.

History of WKC [Internet]. WKC; [accessed on-line November 16, 2008]. Available from http://www.wkc-solingen.de/newshop/history.html.

Kleines Werkzeugmuseum [Internet]. Wolfgang Jordan; [accessed on-line November 20, 2009]. Available from http://www.holzwerken.de/museum/index.phtml

Correspondence:

Balcar, Jan: email correspondence to the author dated May 15, 2004.

Blazicek, Andrej: email correspondence to the author dated April 20, 2004.

Clarin, Ron: email correspondence to the author dated April 13, 2009.

Cutler, Bob: letter to the author dated November 17, 2009.

Huddle, Kenneth: email correspondence to the author dated December 3, 2009 and February 25, 2010.

Jensen, Robert: letter to the author dated October 11, 2009 and December 8, 2009.

Mathisen, Per: email correspondence to the author dated October 18, 2001.

Shomper: letter to the author dated November 16, 2009.

Speed, Jon: email correspondence with the author dated November 20, 2009, December 03, 2009 and December 5, 2009.

Sturgess, Geoffrey: email correspondence to the author dated December 3, 2009 and January 7, 2010, January 18, 2010. Letter to the author dated May 3, 2010.

Veselá, Zuzana: email correspondence to the author dated May 11, 2010.